PHYSICS THROUGH THE 1990s

Atomic, Molecular, and Optical Physics

Panel on Atomic, Molecular, and Optical Physics

Physics Survey Committee

Board on Physics and Astronomy

Commission on Physical Sciences, Mathematics, and Resources

National Research Council

NATIONAL ACADEMY PRESS
Washington, D.C.   1986

NATIONAL ACADEMY PRESS 2101 Constitution Avenue, NW Washington, DC 20418

NOTICE: The project that is the subject of this report was approved by the Governing Board of the National Research Council, whose members are drawn from the councils of the National Academy of Sciences, the National Academy of Engineering, and the Institute of Medicine. The members of the committee responsible for the report were chosen for their special competences and with regard for appropriate balance.

This report has been reviewed by a group other than the authors according to procedures approved by a Report Review Committee consisting of members of the National Academy of Sciences, the National Academy of Engineering, and the Institute of Medicine.

The National Research Council was established by the National Academy of Sciences in 1916 to associate the broad community of science and technology with the Academy's purposes of furthering knowledge and of advising the federal government. The Council operates in accordance with general policies determined by the Academy under the authority of its congressional charter of 1863, which establishes the Academy as a private, nonprofit, self-governing membership corporation. The Council has become the principal operating agency of both the National Academy of Sciences and the National Academy of Engineering in the conduct of their services to the government, the public, and the scientific and engineering communities. It is administered jointly by both Academies and the Institute of Medicine. The National Academy of Engineering and the Institute of Medicine were established in 1964 and 1970, respectively, under the charter of the National Academy of Sciences.

The Board on Physics and Astronomy is pleased to acknowledge generous support for the Physics Survey from the Department of Energy, the National Science Foundation, the Department of Defense, the National Aeronautics and Space Administration, the Department of Commerce, the American Physical Society, Coherent (Laser Products Division), General Electric Company, General Motors Foundation, and International Business Machines Corporation.

Library of Congress Cataloging in Publication Data
Main entry under title:

Atomic, molecular, and optical physics.

(Physics through the 1990s)
Bibliography: p.
Includes index.
1. Physics—Research—United States. 2. Atoms—Research—United States. 3. Molecules—Research—United States. 4. Optics—Research—United States. 5. National Research Council (U.S.). Panel on Atomic, Molecular, and Optical Physics. I. National Research Council (U.S.). Panel on Atomic, Molecular, and Optical Physics. II. Series.
QC44.A86   1985   530'.072073   85-11524

ISBN 0-309-03575-9

First Printing, April 1986
Second Printing, July 1986
Third Printing, November 1986
Fourth Printing, November 1987

Printed in the United States of America

Preface

This report was prepared by the Panel on Atomic, Molecular, and Optical (AMO) Physics of the Physics Survey Committee in response to its charge to describe the field, to characterize the recent advances, and to identify the current frontiers of research. In carrying out this task, we were helped immeasurably by the members of the AMO community and others whose names appear on the following pages. We thank all of them for their contributions.

Acknowledgments

In preparing this report we have been generously assisted by many members of the AMO community, by colleagues in neighboring fields, and by the National Research Council's Committee on Atomic and Molecular Science whose preliminary work helped to pave the way. We particularly thank L. Armstrong, Jr., and J. L. Dehmer for their assistance. In addition, we thank the following for their contributions: N. Bardsley, G. W. Baughman, G. B. Benedek, H. G. Berry, R. S. Berry, R. J. Bieniek, G. C. Bjorklund, C. Bordé, R. G. Brewer, S. J. Brodsky, A. Chutjian, B. Crasemann, S. Datz, P. M. Dehmer, R. D. Deslattes, G. H. Dunn, J. H. Eberly, R. R. Freeman, A. C. Gallagher, T. F. Gallagher, T. F. George, H. M. Gibbs, H. R. Griem, J. L. Hall, S. E. Harris, R. J. W. Henry, J. T. Hougen, M. Inokuti, E. P. Ippen, W. R. Johnson, M. Jones, B. R. Junker, W. E. Kaupilla, H. D. Kelly, E. G. Kessler, A. Kupperman, S. R. Leone, S. Liberman, D. R. Lide, Jr., J. C. Light, C. D. Lin, W. C. Lineberger, T. E. Mady, L. Mandel, V. McKoy, E. Merzbacher, W. E. Meyerhof, D. E. Murnick, J. A. Neff, L. J. Neuringer, D. W. Norcross, R. M. Osgood, A. Owyoung, D. E. Post, Jr., D. E. Pritchard, W. P. Reinhardt, J. S. Risley, R. J. Saykally, D. W. Setser, S. J. Sibener, R. E. Smalley, S. J. Smith, W. W. Smith, A. F. Starace, A. Temkin, D. G. Thomas, J. P. Toennies, R. F. C. Vessot, H. Walther, J. C. Weisheit, and D. C. Wineland.

This report was reviewed by the Committee on Atomic and Molecular Science, National Research Council, whose members are F. M. Pipkin, *Chairman*, L. Armstrong, Jr., R. S. Berry, R. Bersohn, J. L. Dehmer, G. H. Dunn, D. E. Golden, G. S. Hurst, H. P. Kelly, K. Kirby, D. Kleppner, J. Macek, G. A. Martin, C. B. Moore, J. T. Moseley, F. Richard, A. Temkin, N. H. Tolk, J. C. Weisheit, and R. C. Woods; D. C. Shapero, *Staff Director.*

In addition, we thank the following for helpful comments and criticism: H. J. Andra, P. L. Bender, R. B. Bernstein, H. G. Berry, N. Bloembergen, J. Brossel, J. M. Deutch, U. Fano, E. N. Fortson, R. Geballe, S. Haroche, W. G. Harter, P. J. Hay, E. J. Heller, V. Hughes, W. A. Klemperer, Y. T. Lee, R. T. Pack, N. F. Ramsey, C. K. Rhodes, B. Schneider, A. E. Siegman, W. C. Stwalley, and W. R. Wadt.

Contents

Summary

THE NATURE OF THE FIELD

The goals of atomic, molecular, and optical physics (AMO physics) are to elucidate the fundamental laws of physics, to understand the structure of matter and how matter evolves at the atomic and molecular levels, to understand light in all its manifestations, and to create new techniques and devices. AMO physics provides theoretical and experimental methods and essential data to neighboring areas of science such as chemistry, astrophysics, condensed-matter physics, plasma physics, surface science, biology, and medicine. It contributes to the national security system and to the nation's programs in fusion, directed energy, and materials research. Lasers and advanced technologies such as optical processing and laser isotope separation have been made possible by discoveries in AMO physics, and the research underlies new industries such as fiber-optics communications and laser-assisted manufacturing. These developments are expected to help the nation to maintain its industrial competitiveness and its military strength in the years to come.

EDUCATIONAL ROLE

AMO physics plays an important role in the education of scientists in the United States at both the undergraduate and graduate levels.

1

University-based AMO research prepares students for careers in basic and applied science in industry, in national laboratories, and in universities. Approximately 140 Ph.D. degrees are awarded each year in AMO physics.

CONTRIBUTIONS TO NATIONAL PROGRAMS

AMO physics contributes broadly to the nation's programs in energy. Experimental and theoretical data from AMO laboratories are needed for fusion research with magnetic or inertial confinement. Spectroscopy and laser scattering are important diagnostic techniques for plasma fusion devices. Inertial-confinement experiments employ charged-particle devices and high-power lasers whose origins lie in AMO research. Using methods from modern optics, the chemistry of combustion can be studied in an engine as it runs, leading to improved efficiency of aircrafts, ships, and automobiles.

Basic research in AMO physics has revolutionized important areas of military technology. Atomic clocks and laser gyroscopes are central to modern navigation and global positioning systems; fiber-optics communication is widely used in ships, tanks, and planes. Data on atomic and molecular processes from AMO laboratories are vital to the understanding of atmospheric and meteorological phenomena that affect military scenarios. Lasers are used for range finding, guidance, optical radar, and numerous other applications; high-power lasers are being employed in new classes of countermeasures and directed energy weapons systems.

AMO research also contributes broadly to the nation's environmental program. Atomic and molecular data from AMO laboratories are crucial to understanding the chemistry of the atmosphere. Remote-sensing methods employing lasers and laser spectroscopy permit pollutants to be monitored at long distances. Much of our understanding of the effect of ionizing radiation on biological systems is based on data and theoretical research from AMO physics.

RECENT ADVANCES IN BASIC ATOMIC, MOLECULAR, AND OPTICAL SCIENCE

AMO physics encompasses broad areas of theoretical and experimental research on matter at the atomic and molecular level and on light. A few of the recent advances in atomic physics include optical spectroscopy of exotic atoms, new tests of quantum electrodynamics through ultraprecise measurements on individual trapped electrons and positrons, the production of very slow, highly charged ions, the prediction and study of spontaneous electron-positron formation in high nuclear fields, and the first direct measurement of dielectronic recombination. In molecular physics the advances include the development of general techniques for studying molecular ions, the creation of clusters (small groups of isolated molecules), surface scattering with supersonic molecular beams, and the discovery of energy localization in polyatomic molecules. Advances in optics include the first direct measurement of the frequency of an optical transition, the development of ultraprecise optical spectroscopy and ultrasensitive detection of atoms and molecules, laser cooling of ions and atoms, the coherent generation of far-ultraviolet light, optical bistability, and the creation of numerous new types of lasers and nonlinear optical techniques. These discoveries and numerous theoretical advances, including new approaches made possible by computers, have combined to make the past decade of AMO physics a period of substantial scientific progress and unprecedented productivity.

RESEARCH OPPORTUNITIES

The field of AMO physics is moving forward rapidly in wide areas of research on the structure and dynamics of atoms and molecules, the control and generation of light, and the fundamental laws of physics. From among the many activities in AMO physics, the Panel has identified a series of topics that hold promise for rapid advance and high scientific reward. These topics form the basis of a Program of Research Initiatives that is described in detail in the report. The initiative in atomic physics includes tests of fundamental physical laws, the development of high-precision techniques, and research on the many-electron problem and on the dynamics of atomic collisions. In molecular physics the research is centered on understanding the motion of electrons and nuclei in molecular fields and the possibility of controlling the exchange of energy and particles during molecular collisions. The initiative in optics includes the development of coherent light sources from the infrared to the x-ray regions, research on new methods of spectroscopy, and quantum optics.

This program is intended to advance our knowledge of basic AMO science, assure that the field can continue to provide essential data and new techniques for the other sciences, and allow AMO physics to continue its contributions to vital national programs and industry. The program is needed to provide the research environment that is essential for the training of professional scientists for careers in industry, in government laboratories, and in universities.

PRIORITIES OF RESEARCH

AMO physics in the United States advances most often by the efforts of scientists working in small groups on highly diverse problems. The research is pursued by over 300 of these groups in universities, in national laboratories, and in industrial laboratories.

The great strength of AMO physics in the United States is due to the high quality of many of these groups. After a decade of severe winnowing, the remaining groups are seriously threatened by under-funding and the lack of equipment. To assure that the scientific opportunities in AMO physics can be pursued in the United States, the first priority must be to assure the continued vitality of the best of these groups and at the same time to create opportunities for young scientists to enter the field.

RECOMMENDATIONS

The major recommendations are for primary support for atomic, molecular, and optical research in the Initiative Areas. Eight of these areas have been identified. As explained in the report, a 4-year program is proposed at the end of which a total of approximately 140 groups will be pursuing new research in the Initiative Areas. This number is not large considering the breadth of the areas, the variety of scientific opportunities in each of them, the total size of the field, and the need for a reasonable number of new scientists to enter the field, perhaps one a year in each area. At the end of the 4-year period the field should achieve an equilibrium operating level where new work can be started as old work is phased out.

The figures are targets to guide the intensity of the overall effort; they are not meant to fix the exact size of individual grants, the precise number and size of the research groups, or the timetable for starting research in each area.

SUPPORT FOR THE RESEARCH INITIATIVES

Additional funds are essential for AMO groups to carry forward research under the Program of Research Initiatives. The funds are required to support graduate students, postdoctoral workers, and other professional scientists; to help restore the seriously decayed infrastructure of shops, technicians, and special services; to purchase equipment at an orderly rate and to maintain it; to support travel and visitors and to allow enough flexibility for groups to pursue new scientific leads without the 2- to 3-year delay that is now usually required for starting new research. To undertake the new research, the base level of support for basic AMO physics needs to be incremented by $7 million per year (1984 dollars) each year for the next 4 years.

INSTRUMENTATION FOR THE PROGRAM OF RESEARCH INITIATIVES

The instrumentation in most AMO laboratories in the United States is now obsolete, and important scientific opportunities are being lost. The situation is becoming grave. New instrumentation must be provided rapidly if the momentum of research is not to be broken.

The increase in base support recommended above is intended to let the research groups replace instruments at an orderly rate and to maintain the instruments, but it is not adequate for re-equipping obsolete laboratories. For this purpose special one-time support is essential. To equip AMO laboratories for the pursuit of the initiatives, a special allocation of $11 million (1984 dollars) for instrumentation should be made available each year for the next 4 years.

THEORY

In contrast to the situation in Europe, Japan, and the Soviet Union, the theoretical atomic community in the United States is small and highly dispersed. There is a critical need to focus the efforts in this country in order to bring the effort up to the level required to guide and interpret the experimental research. The Panel recommends that the agencies invite and support proposals addressing this issue, for example, by creating centers, workshops, or summer schools where students and active theorists could come together for varying periods of time.

ACCESS TO LARGE COMPUTERS

New approaches made possible by large computers are profoundly changing AMO physics, but the lack of computational facilities for theoretical atomic physicists is seriously hindering activity here. On the basis of a survey of potential users, the Panel recommends that over a 4-year period computer time equivalent to one full-time Cray 1 be made available to AMO physicists, supported by high-speed remote-access facilities.

SPECIAL FACILITIES

Accelerator-based atomic physics and research with synchrotron light sources require facilities that are more expensive than those that have been supported in AMO programs in the past. There are compelling scientific opportunities in both of these areas.

The Panel recommends that proposals be invited and supported for the creation of high-charge ion sources and for accelerator upgrades, at an estimated total cost of $12 million. The Panel recommends that insertion devices be supported for existing synchrotron light sources and that substantial access to them be made available to the AMO community. The Panel endorses the construction of next-generation light sources, both VUV and x-ray, and recommends that beam lines be provided for the AMO community.

1

A Program of Research Initiatives

THE NATURE OF THE FIELD

The central goals of atomic, molecular, and optical physics (AMO physics) are to elucidate the fundamental laws of physics, to understand how matter is composed and how it evolves at the atomic and molecular level, to understand the interactions of light with matter, and to create new techniques and devices. AMO physics is part of a scientific bridge that links physics with astronomy, chemistry, aeronomy, and biophysics.

The experimental and theoretical techniques generated by research on atoms, molecules, and light are often taken up by other areas of physics—nuclear, plasma, atmospheric, condensed matter, surface, and high-energy—as well as by other sciences. The data generated by AMO physics, including the precisely determined fundamental constants of nature, are an essential part of the base of knowledge on which all natural science rests. AMO research also extends into wide areas of applied science; for example, it contributes to the national security system and to the nation's energy programs. AMO laboratories have created advanced technologies that have led to the development of new industries. Such industries are vital for assuring that the United States will retain industrial leadership in the face of the increasing international challenge.

AMO physics plays an important role in the education of scientists in

7

the United States, both at the undergraduate and graduate levels. AMO laboratories, most of which are located on the campuses of colleges and universities, train many of the physicists in our national and industrial laboratories. Approximately 140 Ph.D. degrees are awarded each year in AMO physics. Many of these scientists help to carry forward the nation's energy, military, and environmental programs.

AMO physics is a tremendously diverse field, and this diversity is an essential source of its intellectual vitality. The interplay between various streams of research within AMO physics and neighboring fields of science is demonstrated frequently throughout this volume: an experiment on the quantum electrodynamics of electrons and positrons spurs a new technique for making atomic clocks and optical frequency standards; a close connection is discovered between inner-shell processes in energetic atomic collisions and in elementary chemical reactions; laboratory experiments with low-energy ions cause a re-thinking of a basic astrophysical process. The unity of science is manifest throughout the field of AMO physics.

The influence of AMO physics extends into other areas of science and engineering. To cite one example, many outstanding young scientists in this field now work in Departments of Chemistry. This extension is a sign of the widening realization of the power of the methods of AMO physics for understanding broad classes of natural phenomena. Although chemists may discover fresh viewpoints on chemical reactions from the theory of atomic collisions or from experiments with colliding beams of atoms, there remains a vital core to the subject whose approach is that of the physicist: the search for unity and generality in the natural world. The chemist and biologist need the insights of the atomic physicist to elucidate the diversity of the properties of matter and of organisms. This means that a healthy component of physics will remain necessary no matter how widely the achievements of this field are taken up by other sciences and engineering.

This report attempts to provide a balanced description of AMO physics; to portray its role in the nation's programs in basic science, applied science, and technology; to indicate likely areas for scientific advance; and to describe the steps needed to pursue these opportunities.

ORGANIZATION OF THE REPORT

The remainder of this chapter describes the Program of Research Initiatives that is intended to assure that AMO physics will continue to advance as a science and to meet vital national needs. Research initiatives in AMO physics are described in the last three sections of this chapter. Chapter 2 discusses the role of AMO physics in the United States; Chapter 3 presents recommendations for assuring the continued productivity of the field and for implementing the Program of Research Initiatives. The main body of the report summarizes recent scientific activities in atomic, molecular, and optical physics, in Chapters 4, 5, and 6, respectively, and describes the scientific interfaces and applications of AMO physics, in Chapters 7 and 8, respectively.

INTRODUCTION TO THE RESEARCH INITIATIVES

The field of AMO physics contains diverse scientific opportunities. The vitality of the field stems from the pursuit of a wide range of these opportunities. We describe here a Program of Research Initiatives intended to support scientific innovation and to provide an environment for rapid scientific advance. The goals of the Program are

• To advance our understanding of the laws of physics, the structure and behavior of matter, and the interaction of matter with light;
• To assure continued U.S. leadership in the development of new instruments and techniques;
• To provide experimental and theoretical techniques and physical data that are vital to other areas of science, to industry, and to national programs;
• To attract able young men and women to the frontiers of atomic, molecular, and optical physics;
• To train physicists for careers in universities, in national laboratories, and in industrial laboratories.

In preparing the Program of Research Initiatives we are aware that attempts to predict the most promising avenues of scientific advances are likely to miss the most important developments. For instance, if we had met 10 years ago we would have failed to mention, or would have seriously underestimated, many areas of major progress in the past decade: laser cooling of atoms and ions, low-energy highly charged ions, transient molecular states, Rydberg atoms, molecular clusters, four-wave mixing, phase conjugation, and ultrasensitive detection, for example. The list, which could easily be extended, illustrates the point that 10 years ago AMO physics was developing too rapidly to permit a knowledgeable forecast. Today the field appears to be moving even more rapidly. Nevertheless, we believe that the Program of Research Initiatives represents a realistic basis for scientific advance in the near future.

Because AMO physics is such a diverse field, and because so many different areas have a high potential for scientific reward, the Program is necessarily broad. Fortunately, basic research in AMO physics is generally carried out by small groups rather than by massive research teams. Forefront research in a wide range of activities can be carried out by individuals working in programs whose cost is, by the standards of contemporary physics, small.

INITIATIVE IN ATOMIC PHYSICS

Rapid experimental and theoretical advances have opened new frontiers in many areas of atomic physics. We have selected three areas of basic scientific inquiry in which to exploit these new opportunities:

• *Fundamental Tests and High-Precision Techniques*—to use the atom as a laboratory for the study of basic properties of space and time, to test the elementary interactions and symmetries in nature, and to create new techniques for precision measurements.

• *The Many-Electron Atom*—to obtain physical understanding and quantitative descriptions of many-body systems generally through the elucidation of how electron motions are correlated and to extend these concepts to help interpret the dynamical behavior of atoms within molecules and other, more complex, systems.

• *Transient States of Atomic Systems*—to describe qualitatively and quantitatively the physical nature of intermediate, nonstationary states; the exchange of energy, angular momentum, and particles during atomic collisions; and to understand the role of highly correlated intermediate states.

Fundamental Tests and High-Precision Techniques

The styles of physics—the theoretical and experimental techniques of the various fields of physics—differ dramatically, but central to all of physics is the study of the elementary laws of nature. These studies play a conspicuous role in atomic physics, and activity is at a high level. During the past decade, for instance, time-reversal invariance has been tested at new levels of sensitivity through searches for the dipole moment of the neutron; the isotropy of space with respect to the speed of light has been confirmed by laser interferometry to a few parts in 10^{15}; parity violation by the electroweak interaction has been observed in atoms, and the effect of the Earth's gravity on time has been measured using a rocketborne clock with a stability greater than 1 part in 10^{14}. (See Figure 1.1.) These experiments, which involve measurements of extraordinary sensitivity, often generate techniques that find useful applications in other areas of science and industry. For example, the atomic clock, which was developed to test the effect of gravity on time, now plays a key role in radio astronomy with very-long-baseline interferometers. Atomic clocks have also made possible a new type of navigational system, which can determine one's position anywhere on Earth with an accuracy of about 10 m. The high-precision optical interferometry developed in conjunction with the test of the isotropy of space has applications ranging from gravity-wave detection to seismic monitoring.

Sensitive testing of the limits of quantum electrodynamics (QED) is one of the most important tasks in this area of AMO physics. The confrontation between theory and experiment has moved to a level of precision that is unique in the world of physics. The anomalous magnetic moment of the electron, which has been evaluated in one of the most ambitious calculations of theoretical physics, has now been measured to an accuracy of 40 parts in 10^9 in an experiment that employs a single electron or positron confined in an electromagnetic trap. Together with a new measurement of the radiative-energy-level shift (the Lamb shift) to an accuracy of 9 parts in 10^6, this constitutes one of the most demanding experimental tests of theory ever carried out. The results check the convergence of QED in areas not previously examined; this is a crucial test, considering that QED is the prototype for all gauge theories, including the electroweak theory and quantum chromodynamics. Tests of QED have come from the recent observation of the Lamb shift in the electron-muon atom (muonium) and laser spectroscopy of the electron-positron atom (positronium).

Further discussion of activities in this area can be found in Chapter 4 in the section on Elementary Atomic Physics.

Research opportunities include the following:

• *Elementary Structure*—A hundredfold improvement in the measurement of the electron magnetic moment appears to be feasible using new types of single-particle traps. Provided that the theoretical calculations undergo similar progress, this would test QED and charge-parity-time (CPT) invariance with a precision of 1 part in 10^{13}. There are new opportunities to search for the breakdown of time-reversal symmetry with a hundredfold increase in the sensitivity of the search for the neutron's electric dipole moment, to study particle-antiparticle symmetries, and to study electroweak interactions in atoms.

• *QED in Highly Charged Systems*—Highly charged ions with one or a few electrons can now be produced in fast ion beams. Hydrogen-like uranium (uranium with 91 of its 92 electrons removed) has recently

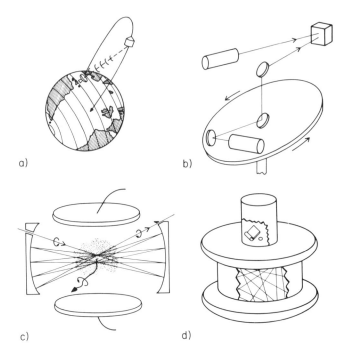

FIGURE 1.1   Four Tests of Fundamental Principles.

a) *Gravity and Time*. According to general relativity, time is slowed by gravity. The effect due to the Earth's gravity is tiny: a clock on the Earth's surface runs slow by only 7 parts in 10^{10} compared with an identical clock in free space. Nevertheless, this effect has been measured to high accuracy by comparing an atomic clock carried by a rocket in a high trajectory orbit with a similar clock on Earth. (Atomic clocks use the natural frequencies of atoms to control the rate of an oscillator.) During the experiment the two clocks maintained a precision of a few parts in 10^{15}, approximately one second in ten million years. Atomic clocks are essential elements of modern navigational and global positioning systems, and they play a vital role in very-long-baseline radioastronomy.

b) *How Constant is the Speed of Light*? Einstein's assertion that the speed of light in empty space is a universal constant, unaffected by any motions of the light's source or by the observer, is accepted as a fundamental law of physics. Like all such laws, however, its validity rests on careful observation. The constancy of the speed of light in different directions has been verified to a few parts in 10^{15} by comparing signals from highly stabilized lasers while their relative directions were changed. Laser stabilization techniques similar to those used in this experiment have applications that range from ultraprecise spectroscopy to new types of metrology and communications.

c) *Parity Violations in Atomic Physics*. The electroweak theory, which unifies the previously separate descriptions of electromagnetism and the weak interaction, is a milestone of modern physics. According to this theory, free atoms can display an intrinsic preference between right- and left-handedness due to what are known as parity-violating interactions. Without the electroweak interactions, parity violations in free atoms would be strictly forbidden. Parity violation effects in atoms have been measured by laboratories in France, Great Britain, the United States, and the Soviet Union. In the experiment illustrated, the parity-violating interactions cause the light been produced. A new area of QED phenomena where both radiative and relativistic effects are large is ripe for study, including research on bound states in strongly relativistic electromagnetic fields, a problem of deep interest in QED and relativistic quantum mechanics.

• *Laser Spectroscopy of Exotic Atoms*—With the coming of age of precision spectroscopy in positronium, accurate measurements of the Lamb shift and elusive relativistic many-body effects in this elementary two-body system have become possible. Muonium has recently been obtained in vacuum, opening the way to spectroscopy of the excited states of muonium. Intense sources of muons and pions are now available, offering the possibility of developing intense pulsed sources of muonium and pionium that are matched to the cycle factor of pulsed lasers. Laser spectroscopy of exotic atoms provides a new arena for the study of quantum electrodynamics in pure leptonic systems.

• *Trapped Electrons, Ions, and Atoms*—Techniques for high-resolution spectroscopy of trapped particles are rapidly advancing. The g factors of electron and positron have been compared to a few parts in 10^{11}; new tests of general and special relativity are under way; and it should be possible to obtain an improved value for the ratio of the electron mass to the proton mass. With trapped-ion techniques it may be possible to compare masses of nuclei to an unprecedented precision, possibly providing a new way to measure the neutrino rest mass. Trapped ion spectroscopy can provide precise tests of theory in fields

---

radiated by laser-excited atoms to become circularly polarized. The experiment has a plane of symmetry, but the circular polarization of the radiated light breaks the symmetry. Although the effects are small—most of the radiated light is linearly polarized and the circular polarization is typically only one part in a million—they have been measured with such precision that the atomic data complement and augment the measurements made by particle physicists using accelerators.

d) *Time-Reversal Symmetry and the Electric Dipole Moment of the Neutron.* The basic laws of physics are ordinarily not sensitive to the direction in which time flows; in general, if time were reversed, the motions would be exactly reversed. In all of physics, only one exception to this time-reversal symmetry has so far been observed; it occurs in the decay of the neutral K meson. Theories of K meson decay predict other breakdowns of time-reversal symmetry. One of the most sensitive tests of this symmetry has been the search for the existence of an electric dipole moment of the neutron. Very slow neutrons from a neutron reactor are trapped in a cell by a valve, and their intrinsic magnetic moment is measured by the technique of radio-frequency magnetic resonance. A strong electric field is applied; an electric dipole moment would reveal itself by a change in the resonance frequency when the electric field is reversed. The experiment illustrated is carried out by a team of physicists using an internationally operated neutron reactor in France. A related experiment is under way in the Soviet Union. The limit on the possible size of the neutron dipole moment has been systematically reduced, placing important constraints on possible theories.

Other applications of AMO physics to basic tests are discussed in Chapter 4 in the section on Elementary Atomic Physics.

ranging from QED and relativistic quantum mechanics in few-electron systems, to electron correlations in multielectron systems. The recent demonstrations of laser techniques for slowing, cooling, and stopping neutral atoms suggest the possibility of also trapping and studying atoms.

Trapped particles can be cooled to temperatures in the millikelvin range by a variety of spectroscopic techniques, opening the possibility of observing unusual states of matter, including Bose condensation for atoms and strongly coupled plasmas for ions. The methods open the way to new types of atomic clocks and optical frequency standards.

Many-Electron Dynamics

The many-electron atom poses a major intellectual challenge for physics. The independent particle model was formulated 50 years ago, and with the power of fast computers it has now been fully realized numerically. However, comparisons with experiments have revealed a rich array of phenomena that cannot be explained by the independent particle model. These result from many-body electron-electron interactions, which give rise to correlated motion in which the electrons drastically affect each other's behavior.

The recognition of the role of dynamical symmetries in correlated systems, and the discovery of correspondences between correlated motion and single-electron motion in strong applied fields, mark recent theoretical advances. There have been numerous experimental advances. Two-electron systems—the "hydrogen atoms" of many-electron systems—can now be studied in new types of high-resolution electron-scattering experiments. By applying laser spectroscopy to relativistic beams of the negative hydrogen ion, the motion of correlated electrons has been observed with a clarity never before possible. Doubly excited states have been discovered in two-electron multiphoton ionization experiments. Using multiple-laser techniques it is possible to create "planetary" atoms—atoms with two very highly excited electrons that can display new types of electronic motion. The spectra of highly excited atoms in strong fields have revealed unexpected systematics that may hold important clues to the behavior of correlated electron systems.

The many experimental and theoretical advances in the study of correlated electron motions make the many-electron atom problem ripe for attack. Such an achievement would have ramifications for our understanding of the structure of matter and would have numerous applications in chemistry and materials science. Furthermore, the study of correlated motions in a relativistic framework is expected to bear on problems in nuclear and particle science.

Further discussions can be found in Chapter 4 in the sections on Atomic Structure, Atomic Dynamics, and Accelerator-Based Atomic Physics.

Research opportunities include the following:

• *Inner-Shell Spectroscopy*—The structure of an atom having an electron missing from an inner shell differs radically from that of a normal atom. In certain cases, correlation effects are enormously magnified by the effects of double-well potentials. Because the energies of inner-shell electron states can be hundreds or thousands of times larger than the energies for the outer electrons, these states are also ideally suited for studies of relativistic and QED effects in heavy atoms and of the basic process of formation and decay of unstable states. Recent developments of high-intensity synchrotron light sources, hard ultraviolet lasers, and charged-particle beams open the way to major advances in this area.

• *Spectroscopy of Highly Charged Ions*—Studies of sequences of ions having the same number of electrons but ever-increasing nuclear charge provide a unique opportunity to investigate the dependence of relativistic and QED effects on nuclear charge. Such investigations yield a better understanding of basic physical theories and an improved knowledge of the properties of the highly charged ions. These ions commonly occur as impurities in hot fusion plasmas, and often they reveal important plasma diagnostic information. In addition, most plausible schemes for short-wavelength lasers involve radiative transitions in highly stripped ions. The recent discovery of methods for creating and trapping highly charged ions provides a major opportunity for high-precision studies of this important class of atomic systems.

• *Multiphoton Processes*—In the intense electromagnetic field of focused laser light, multiphoton processes occur in which an atom or molecule absorbs several light quanta simultaneously. It has been discovered that intense infrared laser pulses can produce doubly charged rare-gas ions in a many-photon absorption process. Two electrons are simultaneously excited in highly correlated intermediate states. Other experiments using ultraviolet radiation have revealed selective multiphoton ionization of inner-shell electrons. The energy exchanged between optical and electronic modes in these processes is greater than any known chemical reaction. Theory is so far lacking, but it appears that multiphoton processes are providing a new avenue of approach to correlated electron dynamics.

• *Highly Excited Atoms*—New techniques for preparing and studying *Rydberg atoms* (atoms with one electron in a highly excited state) allow one to study for the first time regimes in which applied electromagnetic fields are as strong as the electromagnetic fields of the atom. These strong fields have permitted physicists to begin to search for new dynamical symmetries of the atom-field system. Planetary atoms (atoms with two electrons in highly excited states) provide the opportunity to control and study electron correlation phenomena in ways never before possible. The excited electron can serve as a very sensitive probe of the ionic core of the atom and of the quantum-mechanical theory of the coupling between discrete and continuum states.

Transient States of Atomic Systems

Major innovations in the generation, control, and detection of charged- and neutral-particle beams, the creation of new light sources, and new analytical and numerical methods make possible a wide range of precise studies of unstable transient states of atomic systems. Such states are interesting because they are intrinsically nonstationary; the challenge is to describe the *time-dependent* many-electron system and the special time-dependent characteristics of electron correlation and electron-nuclear exchange of energy and momentum. The traditionally sharp distinction between bound and continuum electronic states has largely disappeared as more is learned about the rich structure in the continuum (the sea of unbound states) of atoms, negative and positive ions, and the transient quasi-molecular systems of colliding atomic species. This structure manifests itself in such processes as *autoionization* and *dielectronic recombination*, in electron capture and loss to the continuum, and in *associative ionization* during collisions of excited and ground-state atoms.

Study of the transient electronic states that occur during violent collisions between ions and other ions or atoms has led to the discovery of approximate conservation laws, such as the electron promotion model that was created to explain the unexpected x-ray emission during ion-atom collisions. Other approximate conservation laws govern the evolution of highly excited atoms in electric fields, and a wide range of other atomic-physics phenomena.

The new experimental and theoretical techniques of atomic-collision physics have for the first time permitted the study of the adiabatic limit of ultraslow collisions. These studies are leading to major advances in long-pursued areas such as the threshold law in ionization of atoms by electron impact, quantum effects in cold ion-molecule collisions, and the role of the transient dipole in threshold *photodetachment* of electrons from polar negative molecular ions.

The ability to carry out scattering experiments in which every important variable is measured (often called a complete scattering experiment), complemented by the development of quantitative theories of complex collision phenomena, provides an unprecedented opportunity to the transient states of atomic systems. The goal is to seek hidden symmetries that will help to organize and simplify the description of the dynamics of a large class of complicated multi-electron systems.

Further discussion is presented in Chapter 4 in the sections on Atomic Dynamics and Accelerator-Based Atomic Physics.

Research opportunities include the following:

* *Particle-Beam Collision Measurements*—Advances in the design of ion sources, the cooling and control of ion and neutral beams, ion traps, merged beams, and position-sensitive detectors open the way to definitive studies of excitation, ionization, charge-transfer, and reactive scattering. The results will help to elucidate a variety of processes ranging from chemical rearrangement at ultralow temperatures to the evolution of transient quasi-molecular orbitals in relativistic collisions of stripped ions. Accurate measurements of ionization and recombination rates are needed to understand the radiation from cosmic and laboratory plasmas and charge-transfer reactions, which play important roles in the heating of magnetic fusion plasmas by neutral atomic beams.

New techniques have opened the way to detailed studies of the interactions of atoms and ions with surfaces and solids. The experiments can reveal the dynamics of electron pickup and loss in individual quantum states and open the way to the study of energy transfer under controlled conditions.

* *Collisions in Laser Light*—As tunable lasers are developed over a wider spectral range, and femtosecond pulse techniques become widely available, new opportunities will be created for probing the elementary encounter event in a collision, a process heretofore inaccessible. Such experiments can lead to a new and far deeper understanding of inelastic collision processes and the nature of chemical reactivity. In addition, collisions in laser light can induce atomic

transitions that would not otherwise occur and also reveal new photochemical pathways.

• *Quantitative Collision Theory*—Accurate calculation of collision properties is an essential complement to experimental studies of collision phenomena. Indeed, there are a number of systems for which calculations are more reliable and less expensive than the corresponding experiments, for example, the excitation of ions by electron impact. Major advances in computational methods and machine capacity and capability offer the opportunity for more-ambitious quantitative theoretical studies to deepen our understanding of dynamical processes and to provide essential data for astrophysics, plasma physics, and other applications.

INITIATIVE IN MOLECULAR PHYSICS

Molecular physics is undergoing a renaissance. Triggered by laser methods, advances in molecular beams, and a host of other experimental techniques and motivated by new theoretical developments, molecular physics holds the promise of achieving a deep understanding of fundamental molecular behavior. (See Figure 1.2.) Opportunities abound for major advances, calling for a timely initiative, which we present in terms of two broad areas: the physics of isolated molecules and the physics of molecular collisions.

• *The Physics of Isolated Molecules*—to understand the fundamental bonding and electronic properties of simple molecules; to understand the joint motion of electrons and nuclei in molecular fields; to elucidate the formation, evolution, and decay of excited molecular states; and to investigate novel molecular species.
• *The Physics of Molecular Collisions*—to study the correlated motions of electrons during collisions between atoms and molecules. By using new laser- and molecular-beam methods, one can observe collisions on a quantum state-to-state basis, including the coupling between electric and nuclear motion and the evolution of energy during collisions involving atoms and molecules.

The Physics of Isolated Molecules

Laser methods, synchrotron light sources, modern molecular-beam techniques, and other experimental advances provide the opportunity for a major advance in basic molecular physics and in the host of disciplines that hinge on molecular behavior. It is now possible to prepare virtually any simple molecule in any desired quantum state and to study its structure with clarity. Furthermore, it is possible to study the physics that underlies this structure: the dynamics of electrons moving in molecular fields.

There are two major directions in the research. The first centers on the spectra and structure of simple molecular systems. The study of unusual molecular species, for instance molecular ions, van der Waals and Rydberg molecules, metal cluster molecules and metastable species present new opportunities for understanding molecular structure. One can hope to understand how the electronic properties of molecules evolve into those of bulk material as one progresses from isolated atoms, through dimers and trimers, and to high states of aggregation; to understand what determines the geometric structure of a molecule whose constituents are held together by the weak van der Waals bond, and how the geometry of van der Waals molecules relates to crystal structure; to understand how the level structure of molecular Rydberg states reflects the interactions within the molecular ion core.

The second stream of research emphasizes the dynamical behavior of electrons and nuclei in each others' fields. The electron cloud in an isolated molecule continuously distorts in response to the slow vibrational motion of the nuclei, whereas the nuclei travel on a potential energy surface determined by the electrons' motions: the isolated molecule is a microcosm of elementary chemical behavior. We can now study the motions of electrons in molecules with unprecedented clarity.

Picosecond and femtosecond laser experiments can reveal how energy flows from one part of a molecule to another, the transition to chaotic vibrational motion, and the rates and mechanisms that determine the system's choice of a particular decay mode. Lasers provide highly selective excitation and interrogation schemes that reveal molecular processes at the quantum-state-specific level. The possibilities are enormous; multilaser methods for studying photodissociation, double-resonance spectroscopy, and photoelectron spectroscopy are but a few of the techniques recently demonstrated.

Synchrotron-radiation sources extend and complement the opportunities provided by lasers. These sources give unique access to intense, tunable radiation from the ultraviolet to the hard x-ray range, permitting excitation of virtually any molecular orbital of any stable molecule. Resonances can be observed—shape resonances and autoionizing resonances—in which the molecule is temporarily stalled in a quasi-bound state where the subtle effects of the internal dynamics are enormously amplified.

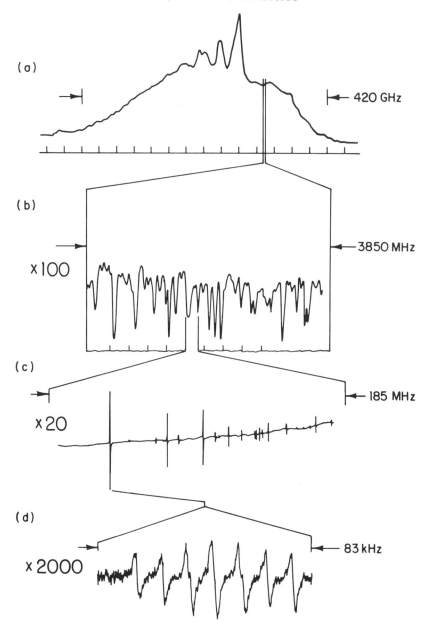

FIGURE 1.2   A Revolution in Spectroscopy. The resolution of molecular spectroscopy has increased by more than one million within a 12-year period. These four spectra of the sulfur hexafluoride molecule show the wealth of new details revealed every time the

The power of electron-scattering spectroscopy has been greatly increased by experimental advances such as spin analysis, coincidence techniques, and position-sensitive detectors. Theoretical progress in understanding the electronic continuum of molecules promises to provide new insight into vibrational and electronic excitation by electron impact, molecular photoionization, and other processes involving highly excited molecular systems.

Further discussion can be found in Chapter 5 in the sections on The New Spectroscopy and Molecular Photoionization and Electron-Molecule Scattering.

Research opportunities include the following:

• *Transient Molecular Species*—Short-lived highly reactive species, including ions, free radicals, and metastable molecules, can now be produced in supersonic beams and studied in detail. These species offer new insights into chemical dynamics and methods for controlling chemical processes.

• *Clusters*—The transition region between a few atoms and molecules and the condensed phase—solid or liquid—is now open to study. Clusters provide a powerful new probe for unfolding the behavior of matter as it evolves from individual atoms to the liquid state, and for studying catalysis. The research bears on our basic understanding of materials and on industrial, technological, and environmental processes.

• *Chaos*—Small molecular systems at low levels of excitation display highly organized internal motions; at higher levels the energy often appears to be distributed randomly. Understanding the transition from orderly to chaotic motion would be an important advance in molecular dynamics and in the quest for state-specific chemical reactions.

---

resolution is improved. The first spectrum, (a), is a conventional infrared absorption spectrum in the region of 10 m. The next spectrum, (b), obtained with diode lasers, shows a blown-up section of a small portion of (a). A rich new structure is revealed. The resolution is limited by Doppler broadening because of the motion of the molecules. In (c) the spectrum was taken by saturation spectroscopy, which avoids Doppler broadening. The resolution of this spectrum is limited by the frequency jitter of the laser. This blowup of a small portion of (b) reveals yet more structure. Finally, in (d) the laser jitter has been reduced by electronic control and the resolution achieves the maximum value allowed by the uncertainty principle. The resolution is limited only by the finite observation time of the molecules. A single sharp line of (c) is revealed to consist of a complex of lines.

The data represent work carried out at laboratories in France and in the United States. Further discussion is in Chapter 5 in the section on The New Spectroscopy. (Courtesy of University of Paris-North, Villetaneuse, France.)

• *Reactive Plasmas*—The physics and chemistry of even simple low-density plasmas is not fully understood. Recently developed laser techniques such as velocity-modulation absorption spectroscopy and laser-frequency-modulated spectroscopy offer powerful new ways to probe the dynamics of such plasmas. One can now measure the quantum-state dependencies of transport properties of ions in plasmas; one can investigate dissociative recombination; one can observe clearly the interaction of excited atoms, molecules, and ions with surfaces. Laser techniques offer an important opportunity to study low-density plasmas, a poorly understood state of matter, and to develop further the rapidly growing industrial use of plasmas.

• *Excited-State Dynamics*—Individual quantum states are now accessible by laser excitation, and their dynamics can be studied in detail. Femtosecond spectroscopy can freeze a molecule as it vibrates, permitting the study of how energy flows within a molecule. By using two or more independently tunable lasers and advanced charged-particle and photon detectors, excited-state photoionization dynamics can be studied with quantum-state specificity at each stage of the process.

• *VUV and X-Ray Photoionization*—Advances in synchrotron radiation have opened the way for studying the fundamental dynamical parameters underlying molecular photoionization, including branching ratios, angular distributions, and spin states. The studies can now be extended to core levels with binding energies in the x-ray range. The energy resolution can be high enough to distinguish between different vibrational modes in polyatomic molecules. By combining lasers with synchrotron radiation sources, a new arena is provided—the study of molecular electron dynamics in core-excited molecules.

The Physics of Molecular Collisions

Experimental and theoretical advances now make it possible to study molecular collisions with the quantum states fully resolved. Moreover, it is now possible to produce molecular beams of a wide variety of unusual molecules—including highly vibrationally excited molecules and molecular ions, radicals, and van der Waals molecules—for use in novel collision studies. Modern lasers can dissociate molecules so as to create a "half-collision" in which dynamical interactions occur only during separation of the products. Such studies now make it possible to study one of the most fundamental aspects of molecular reactions: how the available energy and angular momentum are shared by the reactants. These developments presage a new and deep physical understanding of simple chemical reactions.

In Chapter 5, the sections on Molecular Dynamics and Some Novel Molecular Species contain further discussions.

Research opportunities include the following:

- *State-to-State Chemistry*—With VUV lasers it is now possible to study the chemistry of hydrogen on a state-to-state basis and to compare the results with fully quantum-mechanical calculations of the dynamics on *ab initio* potential surfaces. New state-to-state experiments on heavier systems should open the way to understanding the interplay of translational, vibrational, and rotational energy in simple molecular collisions.
- *Collisions in Laser Light*—The absorption of laser light and the emission of radiation during a collision can drastically alter the course of the collision. This provides a new opportunity for studying the detailed evolution of chemical reactions and may open the way for using lasers to control reaction products.
- *"Half-Collisions"*—In the last decade lasers have been developed that are capable of breaking apart polyatomic molecules with ultraviolet light and multiphoton absorption of infrared visible light. Because the particles fly apart just as if they had initially collided, photodissociation can be viewed as a half-collision, but one in which the reaction complex has precisely known energy and angular momentum. The dynamics of the separating fragments can be studied by modern collision techniques. This research presents a new opportunity to study collision dynamics and to study the evolution of free radicals. The results bear on problems ranging from combustion to atmospheric chemistry.
- *Special Molecules*—Laser techniques now make it possible to produce molecular beams of radicals, of vibrationally hot polyatomics, and of small clusters of most atoms. Many of these are highly reactive. By causing two such beams to cross, the collision dynamics of the two species can now be studied.

INITIATIVE IN OPTICAL PHYSICS

The invention of new lasers and other novel light sources, the development of new methods of spectroscopy and nonlinear optics, and the continued discovery of scientific and practical applications for these new technologies have combined to promote optics and optical physics to a forefront area of contemporary physics. The research has had a strong impact on broad areas of science, on industry, and on many of our national programs. To assure the continued productivity and growth in this area, we propose the following initiatives:

• *New Light Sources*—to develop techniques for producing electromagnetic radiation from the far infrared to the x-ray region, including new lasers and coherent frequency multiplication techniques and new methods for producing short-pulse, short-wavelength radiation.

• *Advanced Spectroscopy*—to develop new methods of ultraprecise spectroscopy, ultrafast spectroscopy, and ultrasensitive detection for the study of atomic and molecular structure and for application to chemistry, materials research, and surface science.

• *Quantum Optics*—to investigate new coherent states of the electromagnetic field, to study nonlinear optical phenomena including the interaction of matter with light at extreme intensities, to exploit new opportunities to study radiative processes of atoms in altered states of the vacuum.

Nonlinear optics, which has been a major arena for scientific and technical advances in optical physics, is not identified in a separate initiative area since it plays a role in nearly every one of the optics initiatives, as well as in the atomic and molecular initiatives.

New Light Sources

During the last decade a host of new light sources has become available for research and for industrial applications and for use in national programs: semiconductor diode lasers whose applications range from high-resolution infrared spectroscopy to fiber-optic communication; the tunable dye laser, which has revolutionized spectroscopy by providing a thousandfold increase in resolution and by opening the way to the preparation and study of atoms, ions, and molecules in states never before achieved; excimer lasers for applications in photochemical processing; the free-electron laser, which holds the promise of providing intense coherent radiation from the infrared through the ultraviolet regions; neodymium/glass lasers that are powerful enough to ignite thermonuclear fusion reactions; and laser-based ultraviolet and x-ray sources. The light sources have opened new areas in the study of the structure of atoms and molecules and in materials science and have applications ranging from ultrasensitive detection of pollutants to metalworking and other manufacturing processes.

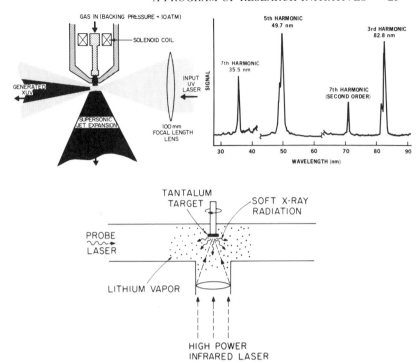

FIGURE 1.3 Short-Wavelength Light Sources. Surface science, chemistry, biology, materials processing, and holography are a few of the many scientific and technical applications that await the development of intense sources of coherent light in the extreme-ultraviolet and soft x-ray regions of the spectrum. The upper left drawing shows a recently developed method for generating extreme-ultraviolet light by harmonic generation of near-ultraviolet laser light in a pulsed supersonic jet of helium. In the spectrum of light from the jet, upper right, signals at the third, fifth, and seventh harmonics of the laser light are clearly visible. The lower drawing illustrates a method for generating a continuous spectrum of radiation in the extreme-ultraviolet and soft x-ray regions. An infrared laser is focused onto a metal target, creating a tiny hot sphere of plasmas, which radiates soft x rays with an efficiency that is often as high as 30 percent. In the setup shown, the x rays are used to photoeject an inner-shell electron from lithium, creating population inversion in the vapor, which may be used to create a short-wavelength laser. (Courtesy of AT&T Bell Laboratories and Stanford University.)

Figure 1.3 illustrates two novel techniques for generating short-wavelength light. Further discussion is presented in Chapter 6 in the section titled Lasers—The Revolution Continues.

Research opportunities include the following:

• *Short-Wavelength Lasers*—New techniques including multiphoton excitation and inner-shell excitation of atoms and molecules offer possibilities for producing laser radiation in the far ultraviolet and in the x-ray regions. Intense beams of pulsed heavy ions can excite a gaseous target and may lead to intense pulsed soft x-ray lasers. Such coherent sources would have an important impact on atomic and molecular spectroscopy, on materials research, on holography, and in many other areas.

• *Nonlinear Processes*—Coherent radiation in the far ultraviolet region can be generated by harmonic multiplication of optical and near-ultraviolet lasers in pulsed gas jets, by four-wave mixing in gases, and by other nonlinear techniques. Such sources are complementary to synchrotron light sources, providing extremely bright light in certain spectral regions with relatively small-scale equipment. New methods for image formation and optical processing based on phase conjugation and other nonlinear processes can have important applications in optical communication, in astronomy, and in the manufacture of integrated circuits.

• *Short Pulses*—19-femtosecond (19×10^{-15} second) light pulses have been generated. As the technology for producing femtosecond pulses improves, increasing numbers of applications can be expected in molecular physics and materials research. Femtosecond pulses make it possible to study the time dependence of basic molecular, solid state, and biological processes. Femtosecond optics provides a new technology for the development of very fast circuits.

• *New Lasers*—New types of lasers are needed, including efficient and powerful optical lasers; simple tunable lasers throughout the infrared, visible, and ultraviolet regions; and ultrastable lasers. Applications for these lasers include communications, trace-element detection and environmental monitoring, chemical and plasma diagnostics, medicine, and manufacturing.

Advanced Spectroscopy

Spurred by the development of new lasers and light sources during the past decade, there has been a revolutionary advance in the spectroscopy of atoms, molecules, and materials. The major source of line broadening in conventional spectroscopy—Doppler broadening—has been effectively eliminated by techniques such as saturation spectroscopy and two-photon Doppler-free spectroscopy. By employing stabilized tunable lasers with those techniques, spectroscopic resolution has been improved by a factor ranging between 10^3 and 10^6. (See Figure 1.2). The methods have been applied to problems ranging from the ultraprecise spectroscopy of exotic atoms to determining the barriers to internal rearrangement in phosphorous compounds that provide prototypes for the basic energy reactions in living organisms. Optical-frequency-counting methods have achieved such high precision and range that the definition of the meter has been fundamentally altered: previously the meter was defined in terms of the wavelength of a spectral line of krypton; now it is defined as the distance light travels in a given interval of time. Light-scattering spectroscopy has found numerous applications in condensed-matter physics, particularly in the study of critical phenomena, and in chemistry, biology, engineering, and medicine. New methods of ultrasensitive detection have made it possible to study molecular ions and free radicals, which are important in many chemical reactions; these methods are being applied to industrial processing and remote sensing. Techniques based on coherent Raman processes have made it possible to study chemical species in hostile environments, for instance in the combustion chamber of an engine. Multiphoton spectroscopy has made it possible to study new classes of atomic and molecular states; while techniques of labeling spectroscopy have vastly simplified the interpretation of complex molecular spectra and have made it possible to prepare molecules in selected quantum states.

Further discussion can be found in Chapter 6 in the section on Laser Spectroscopy.

Research opportunities include the following:

• *Ultraprecise Spectroscopy*—Several different techniques are coming together to permit major advances in ultraprecise spectroscopy. These include highly stabilized tunable lasers, methods for trapping ions and atoms, laser cooling of atomic beams and trapped particles, coherent spectroscopic techniques, and optical-frequency-counting methods. In addition to the application to high-precision spectroscopy, including the study of slow dynamical processes in molecules, these advances create opportunities for new types of optical-frequency standards and atomic clocks, with applications in high-precision measurements and optical communications.

• *Doppler-Free Spectroscopy*—Methods for studying optical transitions with a resolution at the limit of the natural line width offer the possibility of major advances in the study of elementary systems such as hydrogen, positronium, and muonium. The techniques are applicable to large classes of atoms and simple molecules.

• *Ultrasensitive Detection*—Intracavity absorption spectroscopy, laser magnetic resonance, photoacoustic detection, resonant multiphoton ionization, and other techniques permit the study of unstable molecular species. In addition, they offer the possibility of new types of trace analysis for scientific and industrial applications. Ultrasensitive optical detection provides a new method for measuring the solar neutrino flux by detecting nuclei produced by solar neutrinos.

Quantum Optics

Understanding the statistical properties of light and the electrodynamics of matter and light is central to understanding the generation and propagation of light, the transmission of information, and many other physical processes.

Further discussion can be found in Chapter 6 in the section on Quantum Optics and Coherence.

Research opportunities in quantum optics include the following:

• *Preparation of Light in Novel Statistical States*—It may be possible to generate new types of light by learning to control the light's statistical properties. Such a development, particularly the creation of light in squeezed states, has potential applications to quantum metrology and to communication.

• *Optical Bistability*—Optical bistable devices provide new opportunities to study the transition from uniform to chaotic motion. By permitting controlled experiments at a very rapid rate, they may provide a useful tool for studying large-scale chaotic motion, such as the onset of turbulence. Optical bistability has important potential applications to optical processing, including new types of optical logic elements.

• *Electrodynamics at Long Wavelengths*—New types of radiative processes can be seen at microwave or millimeter-wave frequencies using highly excited atoms. The evolution from irreversible to reversible motion can be observed, and the basic source of noise in nature—spontaneous emission—can be modified. New types of electrodynamic effects can be observed, and coherence in small atomic systems can be studied.

2

Atomic, Molecular, and Optical Physics in the United States Today

This chapter summarizes the demographics of atomic, molecular, and optical (AMO) physics and outlines some of its contributions to the community of science and to the nation. The concluding section discusses the changing role of the United States in AMO research.

DEMOGRAPHICS OF ATOMIC, MOLECULAR, AND OPTICAL PHYSICS

The most comprehensive recent study of AMO physics in the United States is the Survey by the Committee on Atomic and Molecular Science (CAMS).* We briefly summarize here the major points.

Size of the Field

Based on 2264 returned questionnaires (from an initial mailing of 6000), the Survey estimates that the community of professional scientists actively working in atomic and molecular science in the United States is approximately 3000.

---

*Subcommittee on Atomic and Molecular Survey, NRC Committee on Atomic and Molecular Science, *Survey of Atomic and Molecular Science in the United States, 1980-1981*, National Academy Press, Washington, D.C., 1982.

Employment

Academic institutions	52%
Industrial or corporate research	18%
Federally funded research and development centers	15%
Government laboratories (civilian or military)	11%
Not-for-profit research organizations	4%

Distribution of Effort

Within broad categories, the research effort is distributed as follows:

Structural properties of atoms and molecules	19%
Atomic and molecular collisional interactions	25%
Interactions with radiation	25%
Techniques and instrumentation	12%
Interfaces with other areas of science and technology	19%

The breakdown between experimental and theoretical work is

Primarily experimental	54%
Experimental and theoretical	20%
Primarily theoretical	36%

THE EDUCATIONAL ROLE OF ATOMIC, MOLECULAR, AND OPTICAL PHYSICS

AMO physics plays an active role in educating scientists in the United States at both the undergraduate and graduate levels. Because the field is one of a small number in physics that permit experimental research in a college setting, AMO physicists are frequently sought for teaching positions in colleges. AMO physics often plays a prominent role in undergraduate education in universities because its laboratories are generally located on campus and the research lends itself to participation by students. By providing research opportunities for undergraduate students in colleges and universities, AMO research plays an effective role in attracting capable students to science.

The major educational role of AMO physics, however, is in the training of professional physicists who are qualified to pursue many different careers in science. AMO research is generally carried out in small groups; it is not unusual for a student to execute an entire experiment single-handedly under the direction of a supervisor—constructing the apparatus, taking and analyzing the data, and working out the theory. The research requires experimental skills including mechanical design and construction, high-vacuum techniques, electronics, lasers, electron and optical spectroscopy, charged and neutral particle beams, and computers. Often a student working in AMO physics becomes expert in several of these areas. Furthermore, in AMO physics it is possible for a single person to work actively both in theory and in experiment, providing an unusually versatile capability. These skills are invaluable for careers ranging from basic physics and chemistry to applied science and engineering.

AMO physicists are in demand for positions in national laboratories and in industry. In national laboratories, for instance, a continued supply of AMO physicists is essential for the development of lasers for applications such as underwater communication, isotope separation, satellite tracking, and defense systems. AMO physicists are deeply involved in fusion research and in environmental monitoring programs. AMO physicists are needed by industry in areas of advanced technology such as fiber-optics communications, laser manufacturing, combustion analysis, optical data processing, photochemistry, and materials preparation.

SCIENTIFIC INTERFACES AND APPLICATIONS

AMO physics contributes broadly to neighboring fields of physics and to other areas of science. Some of the contributions take the form of devices and techniques—the panoply of lasers, light-scattering spectroscopy, supersonic molecular-beam methods, clusters, surface-scattering spectroscopy, and spin-polarized quantum fluids, to name a few. Others are at the deepest scientific level, as for instance in astrophysics (see Chapter 7, section on Astrophysics) or at the interface between nuclear and atomic phenomena (see Chapter 7, section on Nuclear Physics). In addition, AMO physics provides atomic and molecular data, which are essential to fields such as plasma physics and atmospheric science.

AMO physics contributes directly to national needs through a host of applications: plasma diagnostics based on AMO physics are essential to our fusion program; fiber-optics systems play a growing role in civilian and military communications; remote sensing is being increasingly employed for monitoring the environment and industrial pro-

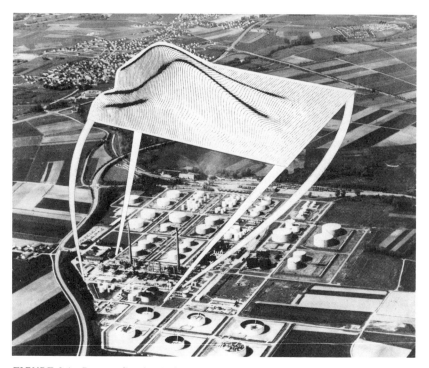

FIGURE 2.1   Remote Sensing Using Lasers. With lasers it is now possible to detect minute traces of chemicals from a distance. In environmental and energy programs the new techniques can be used to detect pollutants in the atmosphere sensitively and rapidly, to study aerosols and smog, to measure turbulence and wind velocity, and to monitor the stratospheric ozone layer. The techniques can also be employed to study combustion in a furnace or in engines while they operate. The illustration shows a blown-up three-dimensional map of the concentration of ethylene glycol that has leaked to the atmosphere from an oil refinery in Germany. The map is superimposed on an aerial photograph of the refinery. (The long arrows illustrate the points on the ground that correspond to the corners of the map.) The gas leak was mapped with a laser 0.5 kilometer away from the plant. The sensitivity of the measurement is 20 parts in 10^9. The peaks in the map reveal two sources of escaping gas. The gas is not coming from the two smokestacks that can be seen in the photo, however; the sources were pinpointed to be two leaks in separate buildings. (Photo courtesy of Max-Planck-Institute for Quantum Optics, Garching, Federal Republic of Germany.)

cesses such as combustion (see Figure 2.1); and laser processing is expected to have a revolutionary impact on manufacturing. AMO physics is vital to innumerable military applications including navigation, communication, and laser-based defense systems; it has contributed to medical care through laser surgery and nuclear magnetic resonance body imaging.

Chapters 7 and 8 describe the scientific interfaces and applications of

AMO research. The activities are broad, and the chapters are by no means comprehensive. Nevertheless, they provide evidence of the many contributions of AMO research to science and to society.

THE ECONOMIC IMPACT OF ATOMIC, MOLECULAR, AND OPTICAL PHYSICS

Over a long period—and sometimes quickly—basic science repays the investment. The return from AMO physics is often large and sometimes rapid. Assessing the full economic impact of AMO physics would be a formidable task, but a few representative examples can help to indicate the magnitude of the return. Nuclear magnetic resonance, whose origins trace back to molecular-beam magnetic resonance, has been applied to a new type of body imaging for medical diagnosis (see Chapter 8, section on Medical Physics). Although the technique is still in its infancy, more than 20 companies are already developing magnetic resonance imaging machines, and hospitals have started to install the devices. The projected sales by 1990 are estimated to be $2 billion to $3 billion. The economic impact of magnetic resonance body imaging is far greater than this, however, for it comes not so much from the sales as from the benefits of improved medical diagnosis: high productivity, better health care, and a better quality of life.

The major economic return from AMO physics in the past decade came from the laser and the developments of modern optics. The conception, development, and reduction to practice of the laser and other modern optical techniques provides a striking illustration of the confluence of academic and industrial research and development in AMO physics. In 1982, the total commercial sales for lasers, laser equipment, and services were $1.5 billion. The laser is having a revolutionary impact on some industries; the fiber-optic industry illustrates how rapidly a new technology can grow. It was not until the mid-1970s, when low-loss fibers were first developed, that commercial applications became a realistic possibility. In 1981 the sales were $24 million; by 1983 they were $620 million. Annual sales in fiber optics in 1990 are projected to be $1.4 billion. The total sales of laser printing equipment through 1988 are projected to be between $5 billion and $10 billion.

The role of lasers and modern optics in industrial applications such as robotics, laser manufacturing, and photochemical processing can be expected to strengthen the nation's economy for years to come. (See Figure 2.2. The use of lasers in manufacturing is described in Chapter 8 in the section on Materials Processing.) These industries are vital to

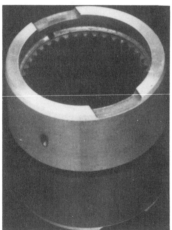 

FIGURE 2.2   Laser-Assisted Manufacturing. Industry is finding more and more uses for lasers, and with the robotics revolution laser-assisted manufacturing promises to become a major industrial force. Laser light is particularly well suited to robotics because the energy can be directed and controlled by computer with unmatched speed and accuracy. Manufacturing processes in which lasers are used include cutting, drilling, welding, surface hardening, and specialized applications such as machining gemstones, ceramics, and semiconductors. Material from the most delicate foil to steel plate 0.5-inch thick can be machined. The upper photograph shows a heavy-duty laser machining installation that cuts, drills, and welds parts in a wide range of geometries. At lower left is an internal gear that was processed by a laser drill. At lower right, the container for a cardiac pacemaker made of titanium is shown being welded by a pulsed laser welder. Further discussion is in Chapter 8 in the section on Manufacturing with Lasers. (Photos courtesy of *Lasers and Applications*.)

the national interest if the United States is to compete successfully with other nations in these times of rapid technological development.

One study of the patterns of future employment in the United States* indicates that by 1990 there will be 200,000 new jobs in the United States in industries related to fiber-optics communications. The use of lasers in manufacturing will generate even more employment: it is predicted that by 1990 the number of these new jobs will be 600,000.

In assessing the economic role of AMO physics, the origin of the laser is worth recalling. The progenitor of the laser was the ammonia beam maser, which was conceived and developed in an academic AMO physics laboratory. Since these early beginnings AMO physics in industrial, academic, and government institutions has made innumerable contributions to the development of lasers and laser-related optical industries. The growth of these industries stands as testimony of the economic return to society that can occur when capable scientists in AMO physics are given the freedom and resources to pursue their goals.

THE HEALTH OF THE FIELD IN THE UNITED STATES

Through the early 1970s, scientific leadership in AMO physics came largely from the United States. Advances such as the invention of magnetic resonance, the discovery of the Lamb shift, the observation of quantum diffraction in high-resolution atomic scattering, and the invention of the laser left no room for doubt about the strength of AMO physics in the United States. The situation is changing. AMO physics was enthusiastically supported in Europe during the past decade, while in the United States it experienced a period of austerity. As a result, the relative level of activity in Europe has advanced dramatically. Europe is now fully competitive with the United States.

In accelerator-based atomic physics, the Europeans and Japanese are likely soon to establish a clear technological advantage. One of the most important new technologies, highly charged ion sources, has been vigorously developed in Europe: the lack of support for developing these sources in the United States makes it difficult for laboratories in this country to pursue research in this scientifically exciting area. If judged by the relative numbers of contributed papers at the International Conference on the Physics of Electronic and Atomic Collisions,

---

Newsweek, Vol. 100, p. 78, Oct. 18, 1982. Based on data from the Bureau of Labor Statistics, Forecasting International, Ltd., and Occupational Forecasting, Inc.

activity in atomic and molecular scattering in Europe has grown conspicuously. In 1971, the breakdown was 47 percent United States versus 36 percent Europe and United Kingdom; in 1983 it was 26 percent versus 54 percent, respectively. Both meetings were held in Europe.

In the early 1970s, there were but a handful of groups at German universities working in optical physics and lasers. Most of the advances in this field came from laboratories in the United States. The rapidly rising number of German publications reveals that this is no longer true. West Germany has assumed a forefront position. In 1974, the Deutsche Forschungsgemeinschaft initiated a program to provide funds for acquisition of modern laser equipment by researchers in German universities. As a result, there were periods when major United States laser manufacturers were shipping more than half of their production to Germany. Most German AMO physics groups now have several state-of-the-art laser systems. In the United States, however, most laboratories could not afford to purchase essential equipment. U.S. laboratories now suffer from a serious lack of lasers. Fifty percent of AMO research involves lasers; the shortage of these devices affects practically every area of AMO research in the United States.

There are many excellent AMO research groups in the United States, and there are numerous opportunities for scientific advance. A central concern, however, is that if one extrapolates the effect of the funding pattern over the past decade into the decade to come, it becomes evident that the quality of AMO research in the United States will be seriously compromised. It is not essential that the United States be preeminent in every area of AMO research, but we must remain competitive and we must maintain excellence in areas where major advances seem likely. We have attempted to define those areas in the Program of Research Initiatives. To assure the continued vitality of AMO physics in the United States it is essential to move forward vigorously on these research initiatives.

3

Recommendations

This chapter recommends action that will permit atomic, molecular, and optical physics (AMO physics) to continue to advance scientifically; to train scientists for industry, government laboratories, and universities; and to contribute to the national programs. The changing pattern of support of AMO physics during the past decade is reviewed in the first section. The following sections describe a plan of action and present recommendations. In the final section the roles of the funding agencies are discussed.

BACKGROUND—THE HISTORY OF SUPPORT

AMO physics in the United States advances by the collective efforts of over 300 groups in academic institutions, in government laboratories, and in not-for-profit institutions. In addition, basic AMO research is carried out in a number of industrial laboratories, including AT&T Bell Laboratories, IBM Research Laboratories, Eastman Kodak Company, and Hughes Research Laboratories.

Four sources provide most of the federal funding for basic AMO research: the Department of Defense (DOD) agencies, the Department of Energy (DOE), the National Science Foundation (NSF), and the National Aeronautics and Space Administration (NASA). The National Bureau of Standards (NBS) also supports AMO physics through its in-house programs and through the Joint Institute for Laboratory

Astrophysics. This diversified pattern of support has intrinsic strengths: it reflects the varied interests within the field and provides flexibility and some protection against sudden changes of policy. Because no agency is responsible for the overall health of the field, however, there is no mechanism for assuring continuity or for responding to major losses of funding.

Until early in the last decade, the DOD agencies provided substantial support, but during the intervening years the DOD largely abandoned its commitment to basic AMO physics. Although the NSF and DOE increased their support, the increase fell far short of the loss. As a result the field has suffered a serious cut in funding.

This history is documented in Figure 3.1 based on the data of Table 3.1. The figures include federal funding of atomic and molecular physics and a small portion of optical physics. The research includes basic AMO physics at universities and university-related private research institutes, plus a fraction of the basic atomic and molecular science at federally funded research and development centers. It is estimated that the figures represent about two thirds of the total support for basic research in AMO physics in universities and university-related research institutes. Table 3.1 and Figure 3.1 provide a realistic portrait of the changing pattern of funding for atomic and molecular physics because the data were assembled over the years using a consistent set of criteria for basic research. Table 3.2 gives the overall funding figures, including optical physics, for 1983.

Comments

• The physics survey* published by the National Academy of Science in 1972 (the Bromley report) pointed out serious problems due to inadequate support that confronted atomic, molecular, and electron physics. Thus the level of funding in 1972, which serves as the starting point in Table 3.1, constitutes a weak base.

• The history of funding between 1972 and 1983 borders on the disastrous. The funding in constant dollars fell by 30 percent. Such a loss by itself would be a serious blow to the field, but the true loss was even greater. The constant dollar figures account for normal inflation, but they do not allow for the increased complexity of science and the need for more-sophisticated equipment. Informed estimates place the

---

*Physics Survey Committee, D. A. Bromley, chairman, *Physics in Perspective*, National Academy of Sciences, Washington, D.C., 1972.

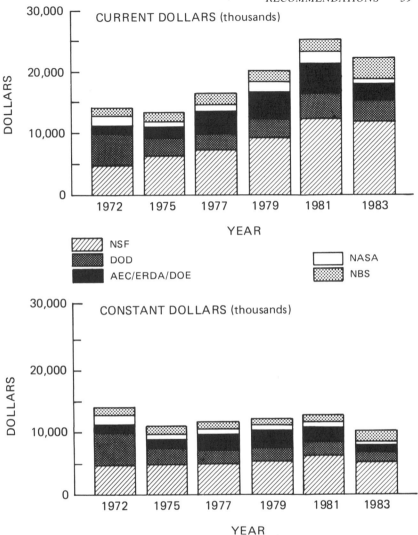

FIGURE 3.1   Funding of basic atomic and molecular physics by the federal government in universities and allied institutes, 1972-1983. Source: Table 3.1.

increase in dollar cost of front-line AMO research over the decade at a factor of 4 or 5. Thus the effective loss to the field is significantly larger than 30 percent; it may be greater than 60 percent. As a result of this loss, the average grant size in AMO physics is now far below the cost of carrying out front-line research.

TABLE 3.1 Number of Grants and Funding of Basic Atomic and Molecular Physics by the Federal Government in Universities and Allied Institutes by Source of Funds in Current and Constant Dollars 1972-1983 (All Dollars in Thousands)[a]

	1972 Current $	1972 No. of grants	1975 Current $	1975 No. of grants	1977 Current $	1977 No. of grants	1979 Current $	1979 No. of grants	1981 Current $	1981 No. of grants	1983 Current $	1983 No. of grants
NSF												
Physics Division	$2,539.8	67	$3,839.5	61	$4,681.3	59	$5,806.7	71	$8,246.6	93	$9,330.0	94
Other Divisions	2,337.8	53	2,561.9	53	2,656.0	55	3,510.8	56	4,404.6	66	2,870.0	38
NSF Total	4,877.6	120	6,401.4	114	7,337.3	114	9,317.5	127	12,651.2	159	12,200.0	132
DOD	5,341.3	115	2,999.1	81	2,844.3	63	3,217.4	61	3,993.7	52	3,410.0	46
AEC/ERDA/DOE	1,108.0	21	1,726.4	31	3,501.9	38	4,273.0	50	4,650.8	48	3,490.0	35
NASA	1,669.0	28	992.0	21	1,388.0	24	1,547.0	30	1,917.0	32	600.0	12
NBS	1,276.0	7	1,601.7	8	1,565.0	2	1,816.7	4	1,991.8	4	2,750.0	9
Total	$14,271.9	291	$13,720.6	255	$16,636.5	241	$20,171.6	272	$25,204.5	295	22,450.0	234
Average project	$49.0		$53.8		$69.0		$74.2		$85.4		$96.0	

Constant Dollars[b] (1972 = 100.0)

	1972	1975	1977	1979	1981	1983
NSF						
Physics Division	$2,539.8	$3,136.8	$3,372.7	$3,626.9	$4,231.2	$4,174.5
Other Divisions	2,337.8	2,093.1	1,913.5	2,192.9	2,259.9	1,284.1
Total NSF	4,877.6	5,229.9	5,286.2	5,819.8	6,491.1	5,458.6
DOD	5,341.3	2,450.2	2,049.2	2,009.6	2,049.1	1,525.7
AEC/ERDA/DOE	1,108.0	1,410.5	2,523.0	2,669.0	2,386.2	1,561.5
NASA	1,669.0	810.5	1,000.0	966.3	983.6	268.5
NBS	1,276.0	1,308.6	1,127.5	1,134.7	1,022.0	1,230.4
Total	$14,271.9 291	$11,209.6 255	$11,986.0 241	$12,599.4 272	$12,932.0 295	$10,044.7 234
Average project	$49.0	$44.0	$49.7	$46.3	$43.8	$42.9

[a] These figures were collected for the Interagency Atomic and Molecular Administrator's Group by the NSF Physics Division. They include federal funding of atomic and molecular physics at universities, university-related private research institutes, and a small fraction of the basic atomic and molecular science at federally funded research and development centers. A portion of optical physics is included. Funding by the National Bureau of Standards (NBS) is principally to the Joint Institute of Laboratory Astrophysics, supported by NBS and the University of Colorado.

[b] Based on the Full Current Cost of Research Price Index System (FCCR), Price Index System Uniform. The FCCR has been developed over an 8-year period utilizing detailed research and overhead expenditures at the University of California, Berkeley; The Ohio State University; and Stanford University with support from the NSF, the American Council on Education, and the National Association of College and University Business Offices.

• The research awards in AMO physics are so small that many experimental and theoretical groups now require two or more awards to sustain an active program. The total number of awards did not, however, increase during the past decade; instead, it slightly decreased. One must conclude that the number of AMO scientists engaged in basic research has declined. The loss of investigators from AMO physics was discussed in the Survey by the Committee on Atomic and Molecular Science (CAMS)* though the Survey found no means to identify the number of investigators who, having lost their support, simply left AMO research. Nonetheless, it is evident that there has been a severe winnowing of the field.

• A conspicuous feature in Table 3.1 and Figure 3.1 is the changing posture of the DOD support. Although DOD support had already dropped significantly in the 5 years preceding 1972, it was still the largest single source of support in 1972, providing 37 percent of the total. By 1983 DOD support had dropped to a relatively small secondary role, providing 15 percent of the total.

• The pressure generated by this severe erosion of support for basic AMO physics has increased to the point that the future vitality of the field is in jeopardy. The problems, which are documented in the CAMS Survey, may be summarized as follows:

The shortage of sustained support for basic AMO physics is making it increasingly difficult for the research groups to maintain the critical level needed for rapid scientific advance; there is little flexibility to move in new directions, to provide adequate support for graduate students, postdoctoral fellows, and visitors. Essential equipment cannot be purchased, and instrumentation in many AMO laboratories is obsolete. The infrastructure of support services—the shops, technicians, and special facilities that are essential for effective research—has seriously deteriorated throughout the United States. There is wide concern that long-term research is being sacrificed for short-term goals. Opportunities in basic research for young scientists have diminished; few universities have the resources to launch the research of a junior faculty member. The pattern of chronic underfunding threatens to stifle the freedom of imagination that is essential for science to flourish.

---

*Subcommittee on Atomic and Molecular Survey, NRC Committee on Atomic and Molecular Science, *Survey of Atomic and Molecular Science in the United States, 1980-1981*, National Academy Press, Washington, D.C., 1982.

TABLE 3.2   Overall Funding of Basic Atomic, Molecular, and Optical Physics by the Federal Government in Universities and Allied Institutes, 1983[a]

Atomic and Molecular Physics	
NSF	
Physics Division	$9,330,000
Other Divisions	2,870,000
DOD	3,410,000
DOE	3,490,000
NBS	600,000
NASA	2,750,000
Total Atomic and Molecular Physics	$22,458.000
Optical Physics	
NSF	$ 891,000
DOD[b]	
ONR	2,650,000
AFOSR	1,000,000
AROD	2,093,000
Total Optical Physics	$ 5,743,000
Total AMO Support	$28,201.000

[a] Data on the support of atomic and molecular physics were compiled using the criteria described in Table 3.1. The support for optical physics is more difficult to identify owing to the lack of a clear distinction between optical physics and optical engineering. The NSF support, $891,000, is estimated as the optical-physics component of a $2.293 million program in quantum electronics in the Division of Electrical, Computer and Systems Engineering. (In addition, approximately $550,000 of the atomic and molecular physics program can be characterized as optical physics.)

[b] The DOD support for optical physics is dominantly the university-based component of research. It does not include support for the free-electron laser ($2.25 million from ONR; $1.5 million from AFOSR), nor does it include costs of basic research in optical physics within the DOD laboratories.

• The panel notes with pleasure that federal priorities for research in the past few years have placed increasing emphasis on the support of basic research. This provides a constructive climate for addressing the serious problems that confront AMO physics.

A PLAN OF ACTION

In order for AMO physics to pursue the opportunities outlined in the Program of Research Initiatives, as well as the other scientific opportunities that are sure to arise, the nation must increase its investment in basic AMO physics. The effects of the pattern of chronic undersupport during the past decade must be countered. In considering alternative routes for accomplishing this, the panel discussed adopting centralized institutional modes and also the possibility of organizing large areas of research around highly visible major facilities. However, the panel concluded that the great strength of AMO physics in the United States has been the high quality of the many diverse, relatively small, research groups. The tradition of scientific innovation and rapid advance by these groups in universities, in government laboratories, and in industrial laboratories, working with the freedom and the resources to pursue scientific leads as they unfold, has been the major factor in the success of AMO physics in this nation. The panel places the highest priority on assuring the continued vitality of these groups.

At present hardly any federally funded AMO university group in the United States has the resources to move forward in new research. We propose a plan to bring the support of a reasonable number of groups to a level where they can undertake new research effectively. The plan is based on the following points:

• A decade of severe underfunding and high turnover has winnowed the field. Attempts to concentrate the existing support on fewer groups would be counterproductive; as much excellent research would be hindered as helped. New support is essential.

• The research community in basic AMO physics in the United States is not large; an estimate based on the CAMS survey indicates that it comprises about 300 individual groups. For AMO physics to retain its vitality, support for a significant number of these groups must be brought to a realistic operating level, and there must be an opportunity for some new groups to start.

• The gap between the available funding and the costs of research has become so great that the timetable for bolstering the research must be prompt.

Based on these considerations, we recommend a 4-year program to allow individual research groups to start work in the areas of the Scientific Initiatives. The goal is to bring the support of the field up to a level where new work can be undertaken as old work is phased out. Special infusions of support would then no longer be needed.

Altogether eight initiative areas are described in Chapter 1. (Three areas are proposed in atomic physics, three in optical physics, and two in molecular physics.) Each of these areas is broad, and one can expect that new opportunities will also occur within the next few years. We propose a plan to start support for approximately four or five groups in each area for 4 years. At the end of the 4-year period there would be approximately 140 groups in the three disciplines. This number is not large considering the breadth of the area, the likelihood of unforeseen scientific opportunities, and the need to start some new groups, perhaps one in each area every year.

The proposed program relates primarily to AMO groups at universities and university-related private research institutes. The needs of AMO groups working in federal laboratories are difficult to document and quantify because much of the basic research is carried out in support of mission-oriented programs. Basic AMO research in federal laboratories is an important component of the field, however, and many of these groups require special infusions of support to move ahead.

RECOMMENDATIONS

Base Support

To undertake new experimental and theoretical research on the Program of Initiatives, funds are required to

—Support graduate students, postdoctoral workers, and other professional scientists;

—Help restore the infrastructure of technicians, shops, and special research facilities that has largely vanished;

—Purchase new equipment at an adequate rate;

—Maintain the equipment;

—Support travel and visitors;

—Allow enough flexibility for groups to pursue new scientific leads without the 2- to 3-year delay that is now often required for starting new research.

The cost of operating a typical active university-based AMO research group is estimated to be about $350,000/year (1984 dollars). Contributing to this figure are the following. The cost to a research grant for supporting one postdoctoral researcher is about $50,000/year. Most experimental groups need a technician, though hardly any AMO group in the United States has been able to retain one; supporting a technician costs a grant typically $70,000/year. Approximately $50,000/year is needed to keep instrumentation up to date. As a result of these and other operating costs, a number of estimates place the average total cost for an experimental group at about $60,000/year for each graduate student in the group.

We estimate that the average increase in funding for an AMO group to move forward effectively is at least $200,000/year (in 1984 dollars). This includes both theoretical and experimental groups, taking cognizance of the higher cost of exprimental work and that the majority of groups are experimental. Based on a target plan of funding for 35 groups a year, the incremental cost for additional funding is $7 million/year for 4 years.

The figures are targets to guide the intensity of the overall effort; they are not meant to fix the exact size of individual grants, the precise number and structure of the research groups, or the timetable for starting research in each area.

Instrumentation

The lack of instrumentation is seriously hindering AMO research in laboratories throughout the United States. During the past decade the complexity of scientific instrumentation increased substantially, causing the costs to skyrocket at the same time that support withered. A state-of-the-art tunable-dye-laser system costs from $100,000 to $150,000; it is not unusual for one experiment to require two of these lasers plus a large amount of peripheral equipment. A modern clean vacuum system for electron scattering costs about $100,000; a state-of-the-art supersonic molecular-beam scattering apparatus costs more. A high quality superconducting magnet and Dewar costs $200,000; a microwave signal synthesizer costs $70,000. Theoretical groups need minicomputers, and some groups need access to larger machines. A university group can require years to acquire funds for a single piece of such equipment.

The DOD-University Research Instrumentation Program, a 5-year program encompassing broad areas of research related to national defense, provides a perspective on the need for instrumentation. The program is funded at $30 million/year; in its first year the requests exceeded $645 million. For the period 1984-1985, 70 requests were submitted by AMO groups; only 10 were funded. These figures demonstrate the overwhelming need for support for instrumentation in AMO physics as well as other areas of university-based research.

Equipment in most AMO laboratories is obsolete. This is true not only for major equipment but for minor equipment such as oscilloscopes and leak detectors. These laboratories must be thoroughly re-equipped in order for the groups to carry out effective research. The increase in base support recommended above is not adequate for re-equipping the AMO laboratories. (It should, however, let the groups replace and maintain instruments as necessary, avoiding this type of instrumentation crisis in the future.) Special one-time support is essential to re-equip AMO laboratories promptly.

The instrumentation required for a typical active university laboratory to start up research initiatives is estimated to cost approximately $300,000. This could purchase, for instance, two laser systems ($220,000), a small vacuum apparatus ($50,000), and smaller equipment including microprocessors, oscilloscopes, and leak detectors ($30,000). Alternatively, the research might require one laser ($150,000) and a superconducting magnet ($100,000) plus smaller equipment, or a fully instrumented state-of-the-art electron-scattering apparatus ($200,000) plus smaller equipment. Some programs will require less than $300,000; but many activities require more-expensive equipment such as specialized magnets, helium liquefiers, or high-power laser systems. This estimate takes cognizance of the fact that theoretical groups require less but that most of the groups are experimental.

Based on a target of re-equipping 35 groups a year, the cost is $11 million/year in 1984 dollars.

Theory

Theoretical atomic physics is not a major activity at most universities in the United States; yet, as demonstrated by activities in Europe, the Soviet Union, and Japan, the field is ripe for exciting and substantial advances. We envision that the general increase in base support will improve the climate for theoretical research, but further steps are required to bring the effort in this country up to the level required to guide and interpret the experimental research. A major problem is the dispersed nature of the theoretical community in the United States. Active theorists are currently found in many different types of institutions, widely separated geographically. There is a critical need to focus efforts in the country, to strengthen the field, and to attract students. The panel recommends that the agencies invite and support proposals addressing this issue, for example, by creating centers, workshops, or summer schools where students and active theorists could come together for varying periods of time.

Access to Large Computers

AMO physics provides a natural testing ground for new modes of description and new mathematical procedures because the underlying physical laws are generally known and precise experiments are possible. New approaches made possible by large computers are profoundly changing AMO physics. For example, numerical integrations of the Schrodinger equation can provide visualizations of time-dependent processes. Threshold laws can sometimes be found by numerical experiments, and the effects of intense radiation fields can be explored numerically. A new order of precision can permit the calculation of the rates of atomic processes that are too difficult or too expensive to measure.

In the last few years the available computing capability in Japan and Western Europe has advanced enormously from powerful computers at individual universities and the construction of efficient networks. The lack of computational facilities for theoretical atomic physicists in the United States has seriously hindered activity here.

The *Report by the Subcommittee on Computational Facilities for Theoretical Research* (National Science Foundation, Washington, D.C., 1981) emphasized the general need for increased computational capability in physics, and the problem is being examined by various bodies. An important component of the need in AMO physics is time on Class 6 computers. The present supercomputer usage is close to the equivalent of half-time on a Cray 1. Based on a survey of potential users, we recommend that over a 4-year period computer time equivalent to at least one full-time Cray 1 be made available to AMO physicists, supported by high-speed remote-access facilities.

Special Facilities

Two areas of AMO physics require special instrumentation or access to large facilities: accelerator-based atomic physics and AMO physics with synchrotron radiation light sources. Because of their relatively high cost on the scale of support for AMO physics, these require separate attention.

ACCELERATOR-BASED ATOMIC PHYSICS

As described in Chapter 4 in the section on Atomic Physics Requiring Larger Facilities, many new scientific opportunities have been created by advances in accelerator-based atomic physics. Pursuing these opportunities requires beam lines at large national-user accelerators, together with accelerator and ion-source installations dedicated to atomic phyiscs. Provision for funding the installation and maintenance of larger facilities, while common in nuclear and high-energy physics, is not traditional within the AMO framework. As a result,

AMO physics in the United States generally operates in a parasitic mode. The number and availability of such arrangements has been diminishing because of cutbacks in other areas, especially in low-energy nuclear physics.

The opportunity exists for the United States to seize a leadership role in accelerator-based atomic physics, but for this to happen it cannot remain a parasite operation on facilities whose major purpose is to serve other fields. In order to exploit this opportunity, the following steps are required:

1. A new generation of sources of highly charged ions needs to be provided for dedicated atomic-physics programs.

2. Dedicated beam lines for atomic physics with fast heavy ions are required at national user facilities.

3. Dedicated atomic collisions accelerator facilities need to be upgraded, taking advantage of recent advances in accelerator technology.

Cost estimates (in 1984 dollars):

1. High-charge ion-source facilities (3)	$3.0 million
2. Beam lines at existing and proposed relativistic heavy-ion accelerators	$4.7 million
3. Accelerator upgrades and replacement	$4.3 million
Total	$12 million

These recommendations are in agreement with the 1981 report of the Committee on Atomic and Molecular Science.*

ATOMIC, MOLECULAR, AND OPTICAL PHYSICS WITH SYNCHROTRON RADIATION

Synchrotron light sources are expected to have a major impact on wide areas of AMO research in the coming decade. Many of the opportunities are summarized in Chapter 4, in the section on Atomic Physics Requiring Larger Facilities; others appear scattered throughout this report.

Toward the end of this decade, the cutting edge of synchrotron-light-intensive research will lie with the new generation of insertion-device

---

*Workshop on Accelerator-Based Atomic and Molecular Sciences, July 1980 (National Academy Press, Washington, D.C., 1981).

machines (using wigglers and undulators). Although conceived and justified for a number of fields and a broad range of applications, AMO physics is a natural component of this user community and stands to gain particularly because of its need for high photon intensity and high resolution. This country has the opportunity to gain world leadership in AMO research using synchrotron light. Existing facilities need to be fully equipped and exploited, and we must be ready with fully instrumented beam lines for atomic and molecular research when the new generation of machines starts operating.

The panel recommends that insertion devices be supported for existing synchrotron light sources and that substantial access to them be made available to the AMO community. The panel endorses the construction of next-generation light sources, both VUV and x-ray, and recommends that beam lines be provided for the AMO community.

RELEVANCE OF ATOMIC, MOLECULAR, AND OPTICAL RESEARCH TO THE FUNDING AGENCIES

Department of Defense

Basic research in AMO physics has revolutionized important areas of military technology. Atomic clocks and laser gyroscopes are central to modern navigation and global positioning systems; fiber-optics communication is widely used in ships, tanks, and planes. Data on atomic and molecular processes from AMO laboratories are vital to the understanding of atmospheric and meteorological phenomena that affect military scenarios. Lasers are used for range finding, guidance, optical radar, and numerous other applications; high-power lasers are being employed in new classes of countermeasures and directed energy weapons systems. Remote-sensing techniques from AMO physics have applications ranging from combustion analysis for engine design to chemical weapons monitoring. Numerous other examples could be cited, but this brief summary should suggest the scope of the impact of basic AMO research on DOD objectives.

The DOD laboratories depend on scientists trained in university-based AMO research. In order for the DOD agencies to identify new opportunities and to avoid technological surprise, and to evaluate proposals involving advanced technologies, the DOD needs to maintain a close relationship with the AMO basic research community.

Considering all of these facts, and considering the enormous contributions to the nation's military program from basic AMO research, it is difficult to understand why the DOD has so drastically reduced its support of basic AMO physics. By abandoning the tradition of long-range support of basic research, guided by scientific imperatives rather than relevance to immediate DOD objectives, the DOD agencies are paving the way to military obsolescence in the coming decades.

This panel recommends that the DOD agencies resume a serious commitment to the support of basic AMO research.

Department of Energy

AMO physics contributes broadly to the support of the mission of DOE to develop existing technology and to discover new technologies for the generation and transfer of energy. The AMO community provides expertise and essential data on electronic and atomic collisions, which are needed to model and to diagnose magnetically confined thermonuclear fusion devices. It provides knowledge of collision processes that are important for isotope separation in the processing of fission fuels and for understanding the interaction of fast fission products with matter. DOE laboratories working on environmental problems depend on the AMO community for scientific talent and for vital data. For example, our understanding of the interaction of radiation with matter—a central problem for the health and environmental mission of DOE—is carried forward at the atomic and molecular level by AMO physicists. The weapons laboratories of DOE require highly skilled scientists to carry out their mission; many of these are trained in university laboratories engaged in basic AMO research.

The problems of future energy technology will require continued expertise and manpower from AMO physics. The vitality of this field over a long period demands vigorous support of basic AMO physics unfettered by immediate program objectives. DOE has attempted to be responsive in this area by supporting high-risk, long-term projects as well as strongly mission-oriented projects. However, the level of support has fallen far behind the costs of research. In order for the AMO community to continue to meet its obligations to our energy programs, the support for basic AMO research within DOE must be brought up to a level that realistically reflects today's needs.

National Science Foundation

The NSF is the agency responsible for federal support of basic physics in the United States. It plays a major role in AMO physics. The NSF Physics Division provides about 33 percent of the total support of U.S. university-based atomic and molecular research; other Divisions provide an additional 17 percent. The NSF relies on the peer review system to judge the quality of the research and to select likely new areas of scientific advance. The Atomic and Molecular Physics Program has attempted to assure the continuity of support for long-range innovative research and to provide grants large enough for effective research. At the same time, the NSF has attempted to provide new starts. Because of budgetary limitations, however, these goals could not be accomplished simultaneously, and the competition for funds has become intense.

In spite of the fact that the NSF is the primary federal agency in the support of AMO physics, it has never acted as the parent agency for the field. The overall size and the scope of the NSF programs reflects a pattern established several decades ago when the DOD agencies provided most of the support and NASA participated more actively than today. Although the atomic and molecular program grew somewhat in comparison with the total support by the Physics Division during the past decade, the increase fell far short of the losses that occurred because of withdrawal of DOD and other agency support. Thus the overall size of the AMO program in the NSF still reflects the situation when AMO physics had broadly based support, rather than the situation today where much of the AMO research is subject to the vagaries of mission-oriented agencies. Furthermore, the average size of the grants has fallen far below the costs of the research.

National Aeronautics and Space Administration

The space missions conducted by NASA yield data whose interpretation requires an understanding of the extensive range of atomic, molecular, and optical processes that occur in atmospheric and astrophysical environments. A new generation of instruments will soon start to extend enormously the sensitivity and spectral range of astronomical observations. From the proposed Shuttle Infrared Telescope Facility, Solar Optical Telescope, 25-meter Millimeter Wave Radio Telescope, and Space Telescope will follow a vast array of data that will create new demands for reliable data on atomic and molecular spectra and atomic and molecular processes. In order to exploit the full power of these new instruments, it is essential to maintain a vigorous experimental and theoretical effort to understand the underlying atomic and molecular processes.

The panel urges NASA to recognize its responsibility to contribute to the support of research in the basic AMO physics involved in atmospheric and astrophysical phenomena.

4

Atomic Physics

The remainder of this report is devoted to an overview of contemporary atomic, molecular, and optical physics (AMO physics). This and the following two chapters survey each of these topics in turn. The final two chapters describe some scientific interfaces of AMO physics and a few of its applications.

The field of atomic physics encompasses three major streams of research. One deals with studies of the elementary laws of nature. This research often involves high-precision measurements—many of the precision measurement techniques of modern science have come from it. The second stream of research is devoted to understanding the structure of atoms and how atoms interact with light. The third stream involves dynamical processes—how atoms interact with electrons, atoms, and ions. This research merges naturally into molecular physics; examples can be found in both this chapter and the next.

ELEMENTARY ATOMIC PHYSICS

Research in elementary atomic physics has flowered during the last decade. Problems include the limits of quantum electrodynamics, the nature of fundamental symmetries and invariance principles, parity-violating interactions, the isotropy of space, the foundations of quantum mechanics, and the effect of gravity on time (see Figure 1.1). Some

53

of these studies are carried out at a level of precision that is unrivaled in modern science.

Much of the research in this branch of atomic physics involves the study of elementary atomic systems, a category that encompasses hydrogen and hydrogenlike atoms, the leptonic atoms muonium and positronium, and a few elementary particles—the electron, the positron, the proton, and the neutron. The subject has obvious overlaps with nuclear and high-energy physics: our criterion for inclusion in AMO physics is more the identity of the observer than the identity of the system. Thus we include studies of the neutral current interaction in atoms and the search for the electric dipole moment of the neutron, carried out by atomic and molecular physicists, but not a search for the proton-antiproton atom carried out by particle physicists. A conspicuous goal of this research is to push to extremes the theoretical and experimental limits on quantum electrodynamics (QED). Nowhere else in physics is the confrontation between theory and experiment so relentless. In this field an accuracy of 1 part in 10^6 is not unusual, and one notable problem, the anomalous moment of the electron, is currently being fought in the eighth decimal place.

Advances in Quantum Electrodynamics

QED is one of the most successful theories ever developed in physics. It is successful in describing nature over a range of lengths spanning 25 decades, from subnuclear dimensions, 10^{-16} cm, to distances as large as 10^9 cm, where satellite measurements have verified the cubic power law falloff of the Earth's magnetic field. Many theories are patterned after QED; its study has been one of the most rewarding pursuits of modern theoretical physics. Atomic physics provided the first experimental evidence for QED, and it continues to provide the most demanding tests of the theory.

Two theoretical advances of the past decade are the discovery of how to combine QED with the weak interaction to create what is called the electroweak theory and the creation of the theory of strong interactions known as quantum chromodynamics. The electroweak theory and quantum chromodynamics belong to a class called gauge theories. Much of the intuition and many of the theoretical techniques for generating gauge theories have come from QED. Confirming where QED is valid, and where it is not, is crucial to understanding this important class of theories.

One troubling problem lies at the core of QED: the calculated corrections to the electron mass or charge are divergent. QED avoids this difficulty by a renormalization procedure in which nonconvergent quantities such as mass and charge are replaced by their experimental values. Calculations are carried out using perturbation theory, essentially an expansion in a power series of the fine-structure constant (α = 1/137). Although the theory appears to work well, it is not known whether the series ultimately converges.

The most exacting tests of QED have come from measurements of the anomalous magnetic moment of the electron and the Lamb shift of hydrogen. Experiments on these during the past decade are among the triumphs of contemporary experimental physics. Under their impetus the theory has been carried forward in one of the most elaborate calculations of contemporary theoretical physics. This confrontation between experiment and theory is unique in all of physics.

A major challenge to elementary-particle physics is to understand quark-antiquark bound-state systems. These have a close analog in QED—positronium. Precise solutions of the two-body relativistic problem are still lacking. New experiments on the structure of positronium may offer valuable clues to the theory.

Magnetic Moment of the Electron and Positron

The first measurement of the electron's magnetic moment anomaly (the departure of the g factor from the Dirac value, exactly 2) had a precision of about 1 percent; today we know it to 40 parts in 10^9 (see Figure 4.1). This astonishing accuracy is a result of the discovery that a single electron can be isolated in an electromagnetic trap and studied for periods up to hours under conditions of almost complete isolation. Spin resonance and various other motions are detected by the interaction of the electron with a tuned circuit. The same interaction can be used to cool the electron to such a low temperature that it is nearly at rest in the trap. The method works equally well with positrons, and the equality of the electron and positron magnetic moments has been confirmed to 5 parts in 10^{11}.

The calculation of the electron magnetic moment anomaly to a precision comparable with the experimental precision is one of the most demanding tests of QED, and one of the most demanding calculations ever made in physics. Computers were used extensively with symbolic as well as numerical techniques. The calculation requires evaluating 891 Feynman diagrams, many of which involve 10-dimensional numerical integrals.

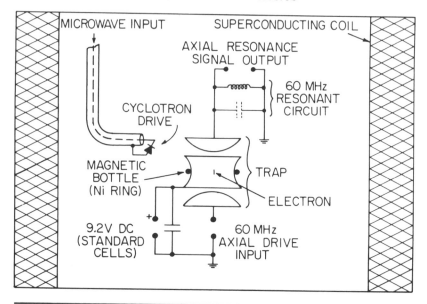

FIGURE 4.1 Studying Quantum Electrodynamics with a Single Electron. One of the most compelling arguments for quantum electrodynamics (QED) was the discovery that the magnetic moment of the electron disagrees with the value predicted by Dirac's theory. The increasingly precise comparison of theory and experiment on this disagreement provides one of the most exacting tests of QED. The disagreement has been measured to 40 parts in 10^9 using this apparatus, which is so sensitive that it employs but one electron at a time. The electron can be trapped and studied for months using a combination of weak electric and strong magnetic fields. Its motion is detected by the tiny current it generates in a circuit attached to the trap. A feeble magnetic "bottle," created by a nickel ring around the trap, allows the spin and cyclotron motions to be detected and compared. The experiment has also been executed with positrons. The electron trap is at the left end of the glass tube. (A vacuum pump is mounted in the right end.) In spite of the deceptively small size of the apparatus, this experiment represents one of the most sensitive and accurate measurements ever made in physics. (Courtesy of the University of Washington.)

Lamb Shift of Hydrogen

The Lamb shift in hydrogen (the small splitting in energy between the $2s$ and $2p$ states) is the second major test of QED. Nature sets a formidable obstacle to the measurement: the $2p$ state is so short lived that the uncertainty principle limits the natural precision in energy of the Lamb shift to 10 percent. Fortunately, the uncertainty principle only sets the scale of difficulty of a measurement—the final precision depends on the experimenter's skill and stamina. A recent experiment employed an intense atomic beam to achieve a resonance line width that was substantially narrower than would be expected from the uncertainty principle at first glance. (The long-lived atoms were preferentially selected.) The final precision was 9 parts in 10^6. At this level of precision the character of the Lamb shift problem changes; the Lamb shift becomes sensitive to the structure of the proton, and hadronic effects must be taken into account. One way out of the dilemma is to think of the Lamb shift as a probe of the proton, thereby testing hadronic physics; another way is to determine the proton structure by high-energy experiments and then combine the high-energy and atomic results to test QED. There is a third alternative: one can avoid all the complexities of hadronic interactions by studying pure leptonic atoms.

Muonium and Positronium

Two species of leptonic atoms are known and both have recently emerged as primary test systems for studying QED. These are muonium (muon-electron) and positronium (positron-electron). Because these atoms decay in a microsecond or less it is a formidable task to satisfy the demands of high-precision spectroscopy for very low energy (the leptons are invariably created at high energy) and a nonperturbative environment, preferably empty space. The annihilation lifetime, hyperfine separation, and Lamb shift of positronium have now been measured precisely; all three effects have provided rigorous tests of theory.

Muonium production has been revolutionized by the creation of meson factories, high-intensity proton accelerators such as the Los Alamos Meson Physics Facility (LAMPF), SIN in Zurich, and TRIUMF in Vancouver. Using a high-intensity muon source, the hyperfine separation of muonium has recently been measured to 3 parts in 10^8. This result now stands as a primary test for QED and the muon's behavior as a heavy electron. By way of contrast, the hyperfine splitting of hydrogen fails as a test of QED at the level of 1 part in 10^6 because of hadronic effects, notwithstanding that it has been measured more precisely than 1 part in 10^{12}.

The discovery of how to make thermal positronium in a vacuum has promoted positronium to a forefront position as a test of elementary theory. Ever since positronium was discovered there has been intense interest in studying its optical spectrum, and optical spectroscopy has recently been successful. By combining the new positronium techniques with methods of modern laser spectroscopy, the 1*s*-2*s* transition has been measured. Because the two bodies have equal mass, positronium offers a stringent test of two-body relativistic theory.

Progress in the optical measurements has been very rapid indeed, and in just about 1 year the accuracy of the optical spectra of positronium has reached the few megahertz level achieved in hydrogen several years ago.

Muonic and Hadronic Atoms

Muonic and hadronic atoms are those in which a negative muon or hadron (π^-, K^-, Σ^-, $\bar{p}$, etc.) is bound to a nucleus by the Coulomb interaction. The spectroscopy of these atoms has been studied by observing their spontaneous emissions, which occur in the x-ray and gamma-ray regions. These data yield values for the hadron's mass and magnetic moment as well as properties of the nuclei. The precision of these measurements was recently increased by employing crystal diffraction spectrometers at the meson factories. For muonic atoms these measurements establish sensitive limits to the mass of the scalar bosons postulated by the electroweak gauge theory.

The most precise value of the muon's basic properties—its mass and magnetic moment—have come from measurements of the Zeeman effect of muonium (the electron-muon atom). The recent discovery of how to make muonium in a vacuum opens the way to a new generation of all of these experiments.

Time-Reversal Symmetry

The origin of the charge conjugation and parity violation (CP violations) observed in the decay of the neutral K meson is one of the great mysteries in physics. If our present understanding is correct, this violation implies a violation of symmetry under time reversal. Despite many careful experiments on a variety of systems, no other violation of time-reversal symmetry has yet been observed.

The most sensitive test for time-reversal symmetry has been a search for an electric dipole moment of the neutron. The original version of this experiment, which used the separated oscillatory field method of molecular-beam resonance, started the experimental search for parity violations in physics, drawing awareness for the first time that parity violations might be expected to occur. Nearly every theory that attempts to explain the CP violation predicts that the neutron will have an electric dipole moment. As the experimental limit has been progressively lowered, many of these theories have been disproved. The sensitivity of the experiment is now incredibly high; it can be compared to confirming the symmetry of a sphere the size of the Earth with a distance less than one tenth the width of the dot in this exclamation point! Nevertheless, a new generation of experiments is under way, and the sensitivity is expected to increase by a factor of 100.

In a series of experiments using atoms and molecules, upper limits have been set on the electric dipole moments of the electron and proton, as well as on possible time-reversal-violating interactions between the electron and nucleus. The ongoing search for electric dipole moments in neutrons, atoms, and molecules provides one of the few possibilities in high- or low-energy physics for solving the CP riddle.

Neutral-Current Parity Violations in Atomic Physics

A key element of the theory that has now unified the electromagnetic with the weak interaction is the prediction that the weak neutral current interactions between electrons and nucleons should produce a parity violation in atoms. The result is that the photons emitted by atoms should "prefer" one circular polarization over the other by a small amount. The effect is extremely small—the wave function is typically distorted by 1 part in 10^{10}. In one class of experiments the parity-violating effect causes the polarization of light to rotate; the required experimental sensitivity is 10^{-7} radian. Observation of these effects is a triumph of experimental ingenuity. Successful experiments have now been carried out with four atomic species: bismuth, cesium, thallium, and lead. These experiments demonstrate neutral-current interactions by low-energy elastic interactions, an arena far distant from high-energy physics where such effects were first observed. The atomic and high-energy experiments are complementary, and at present they are approximately equal in accuracy. Furthermore, the atomic results can be used to put constraints on alternatives to the standard electroweak model, and they provide the opportunities to investigate new classes of phenomena, for instance the possible existence of a second neutral boson, heavier than the recently discovered Z^0 particle. The atomic experiments appear to provide the most sensitive test yet proposed for such a particle.

A molecular-beam magnetic resonance technique, similar to the one used to search for the neutron electric dipole moment, has recently been used to study parity-violating interactions between the neutron and various nuclei. As the neutrons passed through a metal sample a large rotation of the neutron spin due to the weak interaction was observed. The results, which disagreed with the theoretical predictions, have led to a better understanding of nuclear structure. The experimental neutron rotation method constitutes a new tool for examining nuclear structure.

Although the principal tools for studying the electroweak and the strong interaction are those of particle physics, when one contrasts the "table-top" scale of atomic experiments with the scale of high-energy research, it is evident that atomic research is extremely cost effective.

Foundations of Quantum Theory: Is Quantum Mechanics Complete?

Although quantum mechanics is widely recognized as a triumph of twentieth-century thought, persistent questions remain about the validity of the underlying assumptions and the completeness of the theory. The famous debate between Bohr and Einstein attests to the depth of the problem. The fact that Bohr's interpretation is now accepted as the standard model for quantum mechanics does not, of course, preclude the possibility that quantum mechanics may not be able to tell us all that there is to know. Quantum mechanics could be incomplete.

In the mid-1960s, the debate on quantum mechanics was dramatically altered. John S. Bell discovered that if the quantum-mechanical description of phenomena could be supplemented by any further information—including hidden variables that would allow a deterministic interpretation of quantum phenomena—the information could lead to observably different results. Bell showed that correlations in measurements on particles whose initial state was highly correlated would have to lie below a given limit if quantum mechanics were incomplete but that the limit would be somewhat larger if the description by quantum mechanics were complete. The distinction between the two alternatives was presented as an inequality between observables: the completeness of quantum mechanics could be determined by an experimental study of Bell's inequalities.

The experiments involve studies of the correlation in the polarizations of photons successively emitted by a single atom in a cascade of fluorescent steps. The experiments are difficult, and the results were at first ambiguous. Recently, however, Bell's inequalities have been studied in an experiment whose results are clear and decisive. By combining laser-induced fluorescence with modern optical detection methods, the two alternatives could be distinguished with an uncertainty that is about one tenth of the difference. Bell's inequalities were decisively confirmed. The results offer little hope that quantum mechanics can be supplemented by a further description: quantum mechanics appears to be complete.

The limitations of quantum mechanics remain a serious question at the foundations of physics, for the laws of physics grow out of human observations, and there is no reason to believe that they should remain valid in realms where they have never been applied. Nevertheless, Bell's work and the ensuing experiments show that the most obvious possible defect in quantum mechanics is not, in fact, a weakness. The debate will have to turn elsewhere.

Studies of Time and Space

Among the most dramatic research in the general area of high-precision measurement are the studies on phenomena underlying our assumptions about the nature of time and space. Of particular interest are recent experiments on the gravitational red shift and the isotropy of space.

The gravitational red shift refers to the change in the rate of time, or of the frequencies of atomic transitions, due to a gravitational field. The shift has been measured accurately by comparing the rate of a rocketborne hydrogen maser atomic clock with one on the Earth's surface. The red shift was measured with an accuracy of about 2 parts in 10^4. This is a tour de force, for it requires a precision in comparing the two clocks of greater than 1 part in 10^{14}. In addition to confirming the predictions of general relativity, the experiment provided a significant advance in the practical development of atomic clocks and in the art of comparing clocks at large distances.

Fundamental to the theory of special relativity is the assumption that the speed of light in space is a universal constant. In particular, the speed is assumed to be the same in all directions. This has been studied by laser interferometry, and the isotropy of space with respect to the speed of light has been confirmed to within 1 part in 10^8.

These are but two examples from studies that include topics such as the search for cosmological change in the fundamental constants or the comparative evolution of time scales based on different types of atomic clocks. In addition, experimental methods developed in AMO physics are being applied in astrophysics and cosmology. For example, hydrogen maser clocks play vital roles in very-long-baseline radio interferometry, while laser interferometry is at the heart of an important class of gravity-wave detectors.

Future Directions

Measurements of the electron-magnetic-moment anomaly using single trapped electrons are not yet at the fundamental limit. A new version of the experiment has been proposed that may lead to a hundredfold improvement. Provided that the theoretical calculations undergo similar progress, this would test QED and CPT invariance with a precision of 1 in 10^{13}.

Trapped-ion methods are steadily advancing and may lead to a large increase in precision of atomic clocks. Such clocks could provide new tests of general and special relativity. The new trap technology is also expected to provide improved values for the ratio of the electron mass to the proton mass and to provide techniques for comparing the masses of nuclei to unprecedented precision.

Precision spectroscopy of positronium is coming of age. During the next decade physicists can look forward to detailed measurements of the Lamb shift and relativistic effects in positronium. These advances will put increasing pressure on QED theory. The Lamb shift of high-Z hydrogenlike ions has been observed with tunable lasers, and first results have been obtained from potentially more-accurate measurements on this shift in the innermost level. These techniques can be expected to advance to the point where they provide definitive tests of the Z dependence of the Lamb shift, testing essentially different QED contributions.

Muonium has recently been obtained in vacuum, and the Lamb shift has been observed. Intense sources of muons and pions are now available, opening the possibility of developing intense pulsed sources of muonium and pionium that are matched to the duty factor of pulsed lasers. A large and important new field is thus developing—laser spectroscopy of exotic atoms.

ATOMIC STRUCTURE

The elucidation of atomic structure is one of the triumphs of quantum mechanics. The enterprise has been so successful, however, that atomic structure is sometimes regarded essentially as a closed book. This is hardly the case. Consider the following.

—*Trouble with Hydrogen*   The nonrelativistic Schrödinger equation for hydrogen is solved in just about every elementary text on quantum mechanics; it seems unlikely that a serious challenge could remain. Nevertheless, the problem of hydrogen in an applied magnetic field of arbitrary strength is unsolved. Not only are general solutions lacking, for all but the lowest states there are not even any useful approximate solutions. At present, we cannot predict qualitatively how energy levels evolve in regions where the electric and magnetic forces are comparable.

—*The Missing Hamiltonian*   One view of physics is that once the Hamiltonian is known the problem is essentially solved—all that remains is to work out details. Without arguing the pros or cons of this view, we simply mention that the many-body relativistic Hamiltonian is *not* known. The problem of treating many-body systems within the framework of QED poses an important challenge to atomic physics. Even for H$^-$ and He, the simplest two-electron systems, there are no systematic ways to identify the relevant Feynman diagrams. Retardation effects between three or more particles lie at the heart of the difficulty. The problem is more than academic—it is of increasing urgency to astrophysical and plasma problems.

These instances suggest that problems of atomic structure continue to lie in the mainstream of physics. The field is moving forward vigorously, propelled by new spectroscopic techniques and other experimental innovations, and by new theoretical approaches reflecting fresh points of view and increasing computational skill and power.

Loosely Bound Atomic States

The advent of the laser and the development of better sources of negative ions have made it practical to study systems in which one electron is bound loosely. These systems are interesting because it is only near the atomic core that the motion is complex. Over the rest of space the motion of the loosely bound electron is understood precisely.

Rydberg states (highly excited states) of atoms and molecules constitute one class of loosely bound systems (see Figure 4.2). The binding energy can be vanishingly small, and the atoms can be huge. Rydberg atoms with dimensions in the micrometer range have been produced. Energies of Rydberg states can be measured to an accuracy of 1 part in 10^8. In many cases the energy of the states is given by a simple quantum defect formula. The quantum defect is essentially the phase shift in the atomic wave functions due to scattering of the Rydberg electron off the atomic core. The phase shift, which is commonly used in scattering theory, can be measured with spectroscopic precision. This high precision permits the accurate study of spin-orbit, correlation, and relativistic terms that would be far too small to see in conventional scattering experiments.

Negative ions represent a second type of loosely bound system. Because the electron moves in an essentially force-free region, most of the properties of these atomic species were presumed known. When the electrons that are ejected from H^- in stripping reactions were observed, however, there were surprises. If the intensity of ejected electrons is plotted versus energy and angle, forming a two-dimensional surface, only one prominent feature is expected theoretically, a ridge corresponding to a two-body encounter between the loosely bound electron and the target atom or molecule. The experiment revealed a valley that cut across the expected ridge, producing a much more complex electron spectrum. Indeed, observations showed two peaks where only one was expected. It is now known that the double-peaked structure is sensitive to correlations in the ground state of negative ions. This sensitivity to correlation is a new and unexpected feature of the structure of negative-ion continuum states.

Negative molecular ions have also been exploited to investigate rotational and vibrational structure by photodetachment of the loosely bound electron. This structure is rich near the energy thresholds for processes for which the residual molecule is left in an excited state. Recently, systems with two loosely bound electrons have been observed. The motion in such systems is highly correlated and is not amenable to exact treatment. Some features suggest the existence of molecule-like modes in which the two electrons vibrate and rotate about equilibrium positions. This represents a complete breakdown of the conventional independent particle model and signals the onset of collective electron motion.

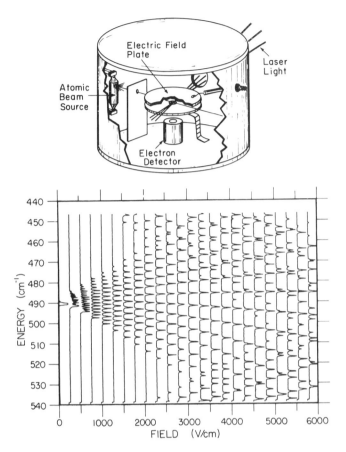

FIGURE 4.2 Atomic Spectroscopy. Lasers have dramatically broadened the scope of atomic physics by allowing the creation and study of new classes of atomic states. This apparatus is used to create highly excited (or Rydberg) atoms. A beam of alkali metal atoms absorbs light from two or three pulsed dye lasers; the atoms are detected by applying a strong electric field that ionizes them. The data show the spectra of Rydberg states of lithium in a series of increasing electric fields. As the energy of the lasers is scanned, the ionization signals are recorded. (They appear as horizontal peaks.) Many different atomic and molecular species have been produced and studied in Rydberg states, as well as species such as planetary atoms in which two electrons are excited. The experiments have stimulated interest in the structure of atoms in strong electric and magnetic fields, a subject that bears on problems in astrophysics, in general dynamics, and in the transition from ordered to chaotic motion. The techniques have also been applied to study collisions, photoionization, superradiance, and electrodynamics. (Courtesy of Massachusetts Institute of Technology.)

Atoms in Strong Fields

That electric and magnetic fields perturb atoms is ancient knowledge; that very strong fields can transform atoms into new species has only recently been discovered. Exhibiting exotic behavior and governed by bizarre scaling laws, these new species are providing unexpected insights into problems such as the nature of the continuum, the effect of symmetry on the structure and dynamics of two- and three-body systems, and the relation between regular and chaotic. motion.

The experiment that launched the subfield was a study of the optical absorption spectrum of high-lying levels of barium in a strong magnetic field. It came as a surprise to discover that the energy levels display periodic structure near the ionization limit and that this structure actually extends into the positive energy region where the electron is free to escape from the nucleus. The levels display the familiar simple structure of a free electron in a magnetic field, except that the level spacing is anomalous. The significance of the anomaly is now appreciated; it is the signal of a new mode of electronic motion, a mode that is characteristic neither of the nuclear electric field nor of the magnetic field alone, but of the joint action of both fields.

In contrast to magnetic fields, which tend to compress electrons close to the nuclei, electric fields tend to tear the electrons out of atoms. One expects that a strong electric field will ionize an atom, but not that it will produce any sort of periodic structure. The discovery of positive-energy periodic structure in an electric field thus came as an additional surprise. In principle, hydrogen has no bound states in an electric field. In practice this difficulty is often overlooked, and the states are treated as if they were stationary. For high-lying states this point of view breaks down, signifying the onset of ionization by tunneling of the electron through the barrier between the potentials of the nucleus and the applied field. The problem has led to new insights into the general relation between discrete states and continuum states.

Most experimental studies with strong electric fields have actually employed alkali metal atoms rather than atomic hydrogen. It seemed reasonable to expect that such an atom will behave like hydrogen if its single valence electron is promoted to a high-lying level, but once again the observations were unexpected. In many cases the energy-level structure is fundamentally different from hydrogen; the field ionization rates may not even remotely resemble the rates calculated by the hydrogenic theory. What was not recognized is the critical role of a special symmetry of hydrogen, a dynamical symmetry, which is destroyed by any departure from a pure $1/r$ potential. (The situation has an analog in planetary motion: any perturbation to the inverse-square-law force causes the orbit to precess.) Our understanding of the role of the dynamical symmetry is now much deeper, including an appreciation of the dramatic consequences of its breakdown.

Double-Well Atomic Potentials

Inner-shell electrons in most atoms are localized radially, confined by an effective potential that has a minimum at each shell's radius. The inner-shell potential is relatively insensitive to the state of the outer electrons, particularly the valence electrons. Consequently, the spectrum for absorption by inner-shell electrons does not vary significantly if one or two of the outer electrons are missing. For some atoms, however, the potential for inner-shell electrons can have *two* radial minima separated by a potential barrier. Usually the electrons are confined in one of the minima, but in some cases they can suddenly switch to the other when the system is perturbed.

One way to perturb the environment of inner electrons is to remove the outer electrons by progressively ionizing the system. Recent measurements of the inner-shell absorption of barium provide a good example of the switching effect. Ba and Ba^+ showed similar features, as generally expected: the features vanished completely in Ba^{2+} in which the electron had switched from the outer well to the inner well, accompanied by a drastic change in the shape of its wave function. Two recent technical advances were crucial to seeing the effect. The first was a bright source of Ba^{2+}: photoionization of Ba^+ by intense laser light made this possible. The second was a continuously tunable source of ultraviolet light. This was provided by a synchrotron light source.

Because an electron in a double-well potential is delicately balanced in one of the two competing states, effects of perturbations are strongly enhanced. Thus, double-well potentials can serve as a "magnifying glass" for studying effects of angular momentum coupling, electron correlations, and relaxation effects.

Collective Atomic States

Our understanding of atomic structure has been dominated by single-particle pictures in which each electron moves independently in an effective potential that is due to the rest of the system. Departures from this idealization are described in terms of correlations, that is, perturbations to single-particle behavior caused by collective motions of two or more particles. This picture continues to serve well for those aspects of atomic structure and dynamics that deal with single-particle excitations. The opening of new spectral ranges by synchrotron light sources and multiphoton absorption of laser light, however, has revealed excited states in which two or more electrons share the excitation energy. These states display a surprisingly wide variation of spectral properties such as decay widths, absorption coefficients, and quantum defects. Attempts to systematize these data have led to the introduction of collective coordinates to describe the highly correlated electron wave functions. It is now realized that for these highly correlated states the independent particle model is not even an adequate zeroth approximation: correlated motion must be faced at the outset.

Among the insights derived from the framework of correlated motion, perhaps the most significant is the prediction of new atomic states, which are entirely collective in nature but which involve only two electrons. Such a state was predicted to occur in the negative hydrogen ion. The ion has only one true bound state but is now known to have infinitely many resonant states that live for 0.1 femtosecond or longer. One of these resonances occurs in a region where conventional independent particle approximations predict no structure. This state was observed spectroscopically in an ingenious and ambitious experiment. A 800-MeV H$^-$ beam at LAMPF was irradiated with laser light. The light was shifted from the visible to the ultraviolet by the Doppler effect. Doppler tuning made it possible to probe the absorption of H$^-$ with great precision. The experiment constituted a major advance in our understanding of highly correlated systems.

A great deal of progress has been made in extending the collective mode description to other systems with two electrons outside of an atomic core. Efforts to describe collective states with three excited electrons are under way. One implication of the collective motion of electrons is that narrow, long-lived states can appear even when the atom has absorbed enough energy to eject two, or even three, electrons. Such states present a new challenge for theory and have prompted searches for new models of atomic structure unrelated to the independent particle picture.

Relativistic and Quantum Electrodynamic Effects in Atoms

In heavy atoms, relativistic effects and effects produced by the interaction between electrons are strongly intertwined. The problem is formidable, for no configuration-space Hamiltonian exists to describe the relativistic interaction between two electrons. Nevertheless, approximate Hamiltonians based on a many-body Dirac equation, with suitably modified instantaneous Coulomb interactions, have been extremely useful in elucidating the interplay of relativity and electron correlation.

The most successful approach to these problems has been the Dirac-Hartree-Fock self-consistent scheme. This treats relativity and electron correlation on an equal footing: neither is considered simply a perturbation to the other. Such an even-handed approach is imperative because of the interplay of the effects. For example, relativistic contraction of energetic inner-shell orbitals changes the potential in which less-energetic, outer-shell electrons move. As a consequence, even though outer-shell electrons are generally nonrelativistic, their energies are greatly affected by relativistic effects.

Experimental searches for certain specific features of atomic processes can aid in understanding the interplay of relativistic and correlation effects. The angular distribution of atomic photoelectrons is one such probe. Without relativistic corrections, the ratios of intensities at different angles is predicted to have a simple form: deviations from this behavior signify the presence of relativistic (as distinct from correlation) effects. The relativistic random-phase approximation was created to study simultaneously relativistic and correlation effects in problems such as this. This method has been used quite successfully for studies of photoionization in xenon and barium.

A serious challenge for atomic theorists lies in the calculation of QED effects in many-electron atoms. Experimentalists will soon be able to measure inner-shell energies of atoms and energies of highly stripped ions with such precision that quantities such as Lamb shifts in many-electron systems can be systematically determined. At present, calculation of these QED effects presents immense difficulties and can only be carried out for simple systems. In addition, even in one-electron systems, calculation of the Lamb shift for very-high-Z elements presents a major challenge.

Work is under way, mainly in Europe, to extend the Dirac-Hartree-Fock theory to molecules. This is needed if one is to treat molecules in which one of the constituent atoms belongs to the sixth or seventh row of the periodic table. Here relativistic effects cannot be ignored, particularly if quantities such as bond lengths are desired that are sensitive to orbital size.

ATOMIC DYNAMICS

Electronic, atomic, and molecular scattering experiments allow us to investigate the transient states of atomic and molecular systems. The experiments are an essential complement to spectroscopy, for spectroscopy generally explores only bound states of systems. Collisions govern the transport of energy in gases and plasmas in environments ranging from high-current switching devices to the magnetically controlled plasma in a tokamak and the atmospheres of hot stars. The applications of atomic collisions research are numerous.

The variety of collision processes is enormous. Fortunately, a number of elementary concepts help to unify atomic-collision phenomena. The long-range nature of the Coulomb field results in a rich and orderly structure of the continuum, manifesting itself as resonances in scattering cross sections. The mechanisms for energy transfer and other state changes during a collision can often be deciphered by comparing the time for the collision (the resonance lifetime) with the characteristic response times of the internal modes: the vibrational period of a molecule or the orbital period of an inner-shell electron. The origin of the large spin-polarization and spin-exchange effects in electronic and atomic collisions can be traced to elementary considerations of the Pauli exclusion principle. Unifying themes such as this provide intellectual coherence to a field that might otherwise seem bewildering.

In this section, we select a few examples to illustrate advances in the field, and we discuss some of the remaining mysteries and new opportunities.

Structure of the Electron Continuum

For electronic energies above the threshold for ionization, or for electron detachment in the case of a negative ion, a continuum of states exists. (See Figure 4.3. A continuum state is the quantum state that describes a free particle. Unlike a bound state, the energy of a continuum state can vary arbitrarily—it is continuous.) The intensity of the continuum states usually varies smoothly with energy, but for these systems the continuum is punctuated by a variety of resonance states in which the electron escapes slowly, as if it were reluctant to depart. These delicate resonant complexes are among the simplest atomic systems involving several simultaneous interactions. Understanding them is a fundamental problem for many-body theory. During the last decade there have been major advances in experimental techniques

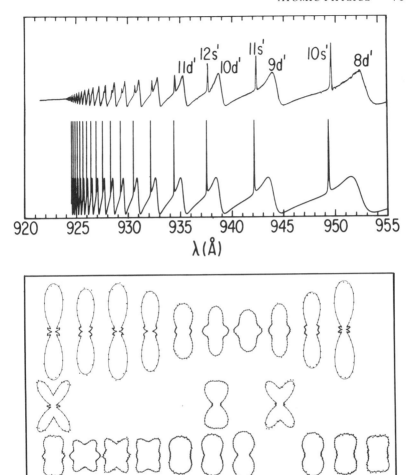

FIGURE 4.3 Photoionization. Photoionization—the process by which an atom or molecule absorbs a photon and ejects an electron—is important whenever a plasma interacts with radiation, as in the interior of stars or in the interstellar medium. The process is important to atmospheric physics, for the Earth's ionosphere is created by photoionization that is due to sunlight. Photoionization is of considerable theoretical interest in atomic and molecular physics because the calculations provide sensitive tests of our understanding of atomic and molecular structure. The upper drawing shows the photoionization cross section of xenon as a function of wavelength. Comparison of the experimental results (upper curve) and theoretical calculations (lower curve) illustrates the excellent agreement between experiment and theory for rare-gas systems. (The overall decrease in the height of the data at short wavelength is due to limited instrumental resolution and is not important.) The lower drawing shows the pattern of photoelectrons (that is, the number of electrons emitted in each direction) from the photoionization of atoms highly excited by laser light. The different shapes of the patterns give precise information about the structure of these atoms. (Courtesy of Notre Dame University and the Joint Institute for Laboratory Astrophysics.)

and the development of new concepts and computational methods. Our understanding of resonance structures and of the role they play in collision processes has advanced enormously.

Until recently, resonance states were classified using a description implicitly based on the independent particle model: each electron in a complex is assumed to move in an average field created by the rest of the system. The simplest of these, the "potential" resonance, is due to the temporary trapping of the scattered electron in a barrier in the potential. A second type, the "core-excited" resonance, occurs when an electron virtually excites a target atom and becomes temporarily attached.

Shortcomings of these traditional classifications have become apparent, and the classifications have now been overturned. For example, the concept of core-excited resonances completely fails to account for structures that were recently observed near the excited states of helium and the alkali metals. The measurements were achieved with new high-resolution electron scattering and negative-ion photodetachment techniques that have revolutionized our ability to study narrow resonance structures. (See Figure 4.4.) These results have forced us to think in terms of a whole new class of electron-atom resonance states. The resonances correspond to unstable multiply excited states of an atomic negative ion. They can occur even if a bound state of the ion is nonexistent, as in the case of He^-. The discovery of these resonances has prompted new approaches to the mathematical physics of excited complexes, for instance the use of hyperspherical coordinates in which two-electron correlations are explicitly built into the representation of the system.

The independent particle model also breaks down under the dominance of correlation effects in the recently discovered Wannier-ridge resonances in He^-. These states were discovered in high-resolution electron scattering. The measured energies suggest two electrons moving in opposition, equidistant from the ion core. The failure of these resonances to fit within the framework of a single-electron Rydberg series clearly indicates an essential three-body structure: two highly correlated electrons tightly bound to an ionic core. Understanding the dynamics of this primary three-body system is an urgent goal for atomic theory: the solution to this problem would be an important step toward the full understanding of the many-body problem.

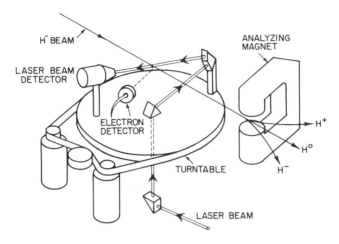

FIGURE 4.4 Spectroscopy with Relativistic Beams. To an ion moving at relativistic speeds, the light from a near-ultraviolet laser can look like it is in the far ultraviolet. This makes it possible to carry out laser spectroscopy in a spectral region where laser light is not available. This principle has been applied to study the H⁻ ion (a proton surrounded by two electrons) with much higher resolution than previously possible. The method combines techniques of particle physics and atomic physics. An 800-MeV H⁻ beam, moving at 85 percent of the speed of light, crosses a beam of laser light. Because of the Doppler effect, the apparent wavelength of the laser light is shortened. As the turntable is rotated, the wavelength "seen" by the H⁻ changes, allowing the ultraviolet spectrum of H⁻ to be scanned. Although H⁻ possesses no excited electronic states, it has many quasi-bound states in which one of the electrons is temporarily trapped by the remaining hydrogen atom. These quasi-bound states, or resonances, are detected by measuring the intensity of emitted electrons. H⁻ is interesting theoretically because it is one of the simplest systems in which the electrons are always highly correlated. (Courtesy of the Los Alamos Meson Physics Facility.)

Dielectronic Recombination

Electron-ion systems display large resonances that are distinctively core excited. These resonances result in a process known as dielectronic recombination.

Dielectronic recombination can occur when an electron hits an ion with slightly less energy than needed to excite the ion. The electron is attracted to the ion until it gets enough kinetic energy to excite one of the valence electrons; at this point the electron becomes trapped in a large highly excited state (a Rydberg level) of the doubly excited neutral atom. If the two excited electrons collide the Rydberg electron can be ejected while the valence electron returns to the ground state of the ion, but if the valence electron radiates its excess energy before such a collision occurs the capture is stabilized and dielectronic recombination is complete. The process occurs much more readily than originally expected. It is so rapid that often it is the leading recombination process in a plasma, governing the equilibrium charge density and dominating the plasma's operating conditions. Consequently, the process is of fundamental interest to the understanding of laboratory, astrophysical, and fusion plasmas.

Until recently dielectronic recombination rates had to be determined indirectly from studying plasma behavior. The situation was unsatisfactory, for the process is so important that precise measurements were essential for verifying the theoretical values. Recently the situation changed dramatically, for within a short period of time three separate groups observed dielectronic recombination using different techniques. Direct measurements have now been made for five ion species by using colliding beams of electrons and ions. For four of the five, the experimental cross sections are *substantially larger* than predicted by theory. Because of the many systems that are affected by dielectronic recombination, including plasma fusion, there is high interest in understanding the source of the discrepancy.

Ultraslow Collisions

When the relative speed of colliding species is small compared with the characteristic speeds of internal motions, the energy levels are slowly perturbed, but the system does not jump discontinuously between states: the system is said to evolve adiabatically. In such a situation the motions of the particles are usually strongly correlated. Adiabatic motion is not normally observed. For instance, when an electron is knocked out of an atom by electron impact, the process is generally nonadiabatic. At an energy just above threshold, however, the two electrons escaping from the ion share a small amount of energy and their motion is essentially adiabatic. This process, near-threshold ionization, provides an ideal system for studying strongly correlated motions.

An early theoretical study of near-threshold ionization of atoms by electrons, using a classical approach, predicted that the cross section would be proportional to the electron energy (measured from the ionization threshold) raised to a rather unlikely power, 1.127. The problem has so far eluded a rigorous quantum solution though an approximate theory predicts that the variation with energy should be close to linear. Results of a careful experiment with ionization in helium agreed with the classical theory. An entirely different experiment was performed at LAMPF. A highly accelerated beam of H^- ions was photoionized by laser light with enough energy to detach the two electrons from the proton. The results can be fit by both the classical and the approximate quantum threshold laws. The adiabatic motion of two electrons near an ion remains an enigma.

Adiabatic motion can also be observed in photodetachment, the process in which a negative ion absorbs a photon and ejects a single electron, leaving a neutral atom. By employing intense high-resolution tunable lasers, threshold phenomena can be studied with a resolution that exceeds that of conventional electron-scattering studies by a factor of 10^6. The number of applications of the technique is large. Photodetachment studies of OH^- at threshold have already opened the way to observing the adiabatic response of a molecule to an electron. Experiments on two-electron photoejection should soon be feasible, providing an important experimental advance on the elusive three-body Coulomb problem.

Collisions with Rydberg Atoms

The experimental art of creating and detecting highly excited atoms (Rydberg atoms) has rapidly developed to the point where a wide range of precisely controlled conditions are realizable, including orbital shapes and matches between energy levels and level spacings. The techniques have opened the way to the study of large classes of collision phenomena and have led to a number of dramatic discoveries. For example, the cross sections for resonant transfer of rotational and vibrational energy of a polar molecule to the electronic excitation of Rydberg atoms has been found to be enormous, up to a thousand times larger than typical molecular collision cross sections. Another discovery occurred in the study of collisions between Rydberg atoms. The energy levels can be shifted or tuned by applying an electric field. When the excited Rydberg level is tuned to be exactly midway between two adjacent levels, an enormous enhancement in the cross section for energy-level-changing collision was found—close to one million times the area of a typical ground-state atom. This enhancement provides another example of the unusual properties of adiabatic motion.

If the principal quantum number n of a Rydberg atom is large enough, the electron's orbit can be so big that the electron and the ion core essentially interact independently with a neutral target particle. This vastly simplifies the problem. Because the kinetic energy of the Rydberg electron is only a few millielectron volts, Rydberg atoms provide a way to study low-energy electron scattering in a regime virtually inaccessible by conventional scattering techniques.

Approximate Conservation Laws

Collisions of many-electron atoms with many-electron targets appear hopelessly complicated. Even the largest computer cannot accurately model a simple helium-helium collision. Broad organizing principles are essential in order to understand such collisions. One of the most fruitful principles to emerge in the past decade is the electron promotion model. Here is one example of this point of view. The simplest molecular ion, H_2^+, has a high degree of symmetry, higher than its geometry suggests, and this symmetry implies the conservation of certain quantum numbers. The surprising feature is that such quantum numbers can be approximately conserved in energetic collisions of many-electron atoms and ions. The conservation rules are central to our understanding of ion-atom collisions. As ionic projectiles approach atomic targets, the electrons of the system move into superpositions of states of the diatomic molecule. This results in a nonstationary state in which the electron distribution oscillates with frequencies characteristic of the transient diatomic molecule. By measuring the angle- and velocity-dependent electron-capture probabilities, the oscillation frequency can be determined. It is unexpectedly high. This high frequency has been shown to imply a new conservation rule based on the symmetry of the simplest molecular ion. In essence, only states allowed by these new rules can be populated. In many-electron systems these correspond to states with many electrons excited. Because the electrons are promoted to higher levels, this model is known as the promotion model.

The most persuasive evidence for the generality of these rules was discovered accidentally. X rays produced by ions striking solids showed an unexpected high-frequency continuum of unknown origin. This band was found to arise from transitions between H_2^+-like orbitals in the transient molecular states of the ion traversing the solid. The radiation was produced by transitions between molecular states that exist for times of the order of 10^{-16} s. Without the approximate conservation of these new quantum numbers, radiation in the x-ray range could not be understood.

The promotion predicts the appearance of energetic electrons and high-frequency x rays, highly anisotropic shapes, and anomalously large ionization cross sections. These have all been confirmed experimentally.

There is a connection between the promotion model and Rydberg states of atoms in electric fields. When the nuclei of an H_2^+ ion separate to large distances, the system behaves like a hydrogen atom in a uniform electric field. The approximate symmetry reveals a close connection between the structure of high Rydberg states and ion-atom collisions. Approximate conservation laws are a recurring theme of atomic physics; whenever they are discovered they help to unify and systematize widely diverse data.

Toward the Complete Scattering Experiment

Recent advances in experimental technology, including position-sensitive detection, polarized beams, improvements in energy resolution, and fast electronics, have made possible a new range of measurements approaching the complete scattering experiments in which every possible quantum number is specified.

The first scattering experiments in which all the quantum-mechanical observables were determined involved exciting the ^1P states of He by electron bombardment. This is a four-body problem, but it is nonetheless simple because the total electron spin of the target is zero both before and after the collision; and there is no nuclear spin. In such a case, electron-spin effects are negligible and there is no hyperfine structure to be considered. By studying the electron-photon angular correlations following the collision, all the excitation amplitudes, and their relative phases, can be obtained. The experimental data yield accurate values for the electron-atom interaction potential, and this provides an accurate check of the approximations needed to carry out *ab initio* calculations. Recently complete scattering experiments have been carried out for the polarized electrons scattered from polarized hydrogen and from xenon.

Comparisons of Positron and Electron Scattering

Although the positron and electron differ only in charge, their scattering behavior can differ enormously. For example, in slow collisions with helium, the electron-scattering cross section is over 100 times that of the positron, though at sufficiently high energies the two scatter identically. Ramsauer minima are evident in positron scattering of He and H_2, though not in electron scattering. The positron- and electron-scattering cross sections for He and H_2 come rapidly into agreement for energies above 125 eV but not for Ne, N_2, and heavier targets. In most cases studied, electron scattering is stronger than positron scattering.

These differences and similarities can be understood qualitatively by recognizing that the long-range polarization interaction, which is important in low-energy scattering, is always attractive and that it is asymptotically identical for positrons and electrons. The short-range interactions, on the other hand, can be very different. For electrons, the exchange interaction partially cancels out the attractive field of the nucleus; for positrons, the exchange interaction vanishes and the short-range field is repulsive.

ACCELERATOR-BASED ATOMIC PHYSICS

There is a body of atomic phenomena that is seen in collisions involving energetic ion-beam methods. This area of research is sometimes called accelerator-based atomic and molecular physics—most of the experiments employ some type of accelerator—though it could equally well be called high-energy atomic and molecular physics, for the scientific interest often focuses on strongly interacting, highly distorted and excited systems. Also included in this classification are studies of accelerator-produced beams of H⁻, high-charged ions and muonic atoms discussed elsewhere in this report.

A fast beam of highly charged ions can collisionally generate systems whose electronic ionization energies and excitation levels are several kiloelectron volts. These collision fragments offer unique opportunities for electron and photon spectroscopy on highly ionized systems that challenge theory in a new domain of few-electron strongly bound atoms. The transient collision system may have a combined nuclear charge that is so high that a new phenomenon occurs—the spontaneous production of an electron-positron pair. Understanding the dynamics of the collision presents a theoretical challenge to deal with the interaction of a fast, highly charged particle with electrons, atoms, or molecules.

The scientific implications of research in accelerator-based atomic physics extend from quantum electrodynamics to molecular and solid-state physics; its practical applications extend from the creation of new sources and detectors to the development of ion-implantation techniques. Progress in this field has been stimulated by dramatic discoveries of new physical phenomena such as continuum electron capture and x-ray transitions in superheavy quasi-molecules and by steady progress in our ability to deal with the dynamics of violently colliding atomic-ionic systems. The examples below illustrate some of these advances.

Atomic Coherence and Out-of-Round Atoms

By studying the angular distribution of decay products from collisions, the shapes of excited atomic states can be determined. For example, when atoms pass through solid foils or reflect from solid surfaces, situations arise where the collision products all spin in the same direction or where the charge clouds are completely aligned. The light that is emitted as the excited states decay is not isotropic; it emerges in some preferred direction at a given instant. Weak internal forces can perturb the shapes of the excited states causing the spin or the charge cloud to precess. This precession produces a sort of searchlight effect in which the intensity of the light that is observed in a particular direction oscillates in time. These oscillations, called quantum beats, arise from the interference of two or more quantum states that are excited simultaneously. The periods of oscillation can be very short, but the time resolution in these experiments now approaches 1 picosecond (10^{-12} s), allowing the beats to be observed.

Measurements of transient nonspherical atoms are most straightforward in electron-scattering experiments, but the concept has far more general applications. For instance, it has solved a long-standing puzzle: Lyman-alpha radiation is copiously emitted when a molecular hydrogen ion breaks into a proton and a hydrogen atom during a collision with a projectile such as helium. This means that the hydrogen atom emerges in an excited state, in contradiction to the conventional model, which predicts that it should be in the ground state. When the shape of the ion during the collision was determined experimentally, the charge distribution was found to have a node, as in the $2p$ atomic state, but the node was oriented randomly with respect to the internuclear axis. The solution to the puzzle follows immediately, for it can be shown that if the node is parallel to the axis the molecular ion must be in a *pi* state, which dissociates to an excited atom, whereas if it is perpendicular it forms a *sigma* state, which dissociates to a ground-state atom. The reason for the copious Lyman-alpha radiation is simply that the elongated charge cloud is formed randomly with respect to the axis. This is exactly opposite to the behavior under photoexcitation in which the relative orientation is fixed because of angular momentum conservation.

Quantum Electrodynamics of Highly Charged Systems

Precise tests of quantum electrodynamics have been made in the regime where particles interact weakly either with each other, as in hydrogen or muonium, or with an external magnetic field, as in measurements of the electron or muon magnetic moment. A less-well-explored subject is high-Z highly ionized atoms, in which the constant that measures the strength of the coupling of the electrons to the nucleus, Z_α, is not a small parameter. Because of the strong Z dependence of the QED effects, the precise tests in loosely bound low-Z atoms provide no quantitative information on the validity of QED in strongly bound high-Z atoms. Thus independent tests of QED and the renormalization prescription for highly relativistic strongly bound electrons are necessary.

Hydrogenlike beams, produced by foil stripping fast heavy-ion beams, have been used to measure the $2s$ Lamb shift in systems with Z as large as 18 (argon). Precision spectroscopy on fast heavy-ion beams has also been used to measure the $2s$ Lamb shift in heliumlike systems. These experiments have stimulated theoretical efforts to deal simultaneously with radiative corrections and interactions of strongly bound electrons, one of the unsolved problems of QED. In addition, the energy of Lyman-alpha photons produced by foil-excited fast-ion beams of iron and chlorine has been measured. High-precision spectroscopy of slow recoil ions is in progress and is expected to reveal the large $1s$ Lamb shift in a high-Z system.

Pair Production in Transient Superheavies

Nature provides no stable species with Z greater than 92, but during close collisions between energetic heavy ions, quasi-molecules are formed whose combined charge can greatly exceed 137. For example, two uranium nuclei at a collision energy near 1 GeV can achieve an effective nuclear charge of 184. Under such extreme conditions the quasi-molecular system is highly relativistic, offering opportunities to study atomic processes in superheavy systems. If the effective value of Z is greater than 173, the K-shell electron binding energy will exceed twice the rest-mass energy of an electron. It is predicted that in this situation a K vacancy can spontaneously decay by creating an electron-positron pair. In other words, if the nuclear charge is large enough, the vacuum becomes unstable. A second type of atomic pair production, caused by the time-varying fields during the collision, has been observed. A surprising structure in the energy spectrum of the positrons has been observed; this may be the result of the anticipated spontaneous decay. In addition to atomic pair production, quasi-molecules offer opportunities to investigate radiative transitions in systems where the transient magnetic field can reach 10^9 times the strength that can be created in laboratory electromagnets.

Inner-Shell Molecular Orbitals and Molecular Orbital X Rays

Inner-shell vacancy production was long believed to proceed mainly via Coulomb ionization. It is now known that this mechanism is frequently eclipsed by inner-shell electron promotion mechanisms: the inner atomic orbitals evolve into molecular orbitals during the collision, from which they undergo transitions at near degeneracies to vacant orbitals whose ultimate evolution is to excited atomic levels. A relativistic treatment is essential, as well as a careful treatment of the many-body aspects of the problem. Considerable progress has been made in understanding the essential individual molecular orbital transition and how outer-shell vacancies are dynamically transformed into inner-shell vacancies. The results are of some practical importance: inner-shell vacancy production plays an important role in processes such as heavy-ion energy deposition in biological materials and also in ion-beam compression of fusion pellets.

X rays have been observed from discrete transitions between molecular orbitals formed during heavy-ion collisions. Radiative rates increase so rapidly with transition energy that this process is actually easier to observe for transitions between inner orbitals than between outer orbitals. Energy spectra and production probabilities for these x rays are sensitive probes for predictions of inner-shell processes within the molecular orbital model.

Charge Transfer

The transfer of an electron in collisions between ions and atoms is one of the most elementary rearrangement processes, and understanding charge exchange is an important step toward understanding complex reaction processes. Over the past decade, we have learned a great deal about electron transfer of both outer- and inner-shell electrons. The prototype charge-transfer problem is the capture of a single-electron by a fast point-charge projectile. The problem might seem well suited to a simple perturbation treatment, but this does not work. In the last few years comprehensive theoretical treatments have emerged. It is now recognized that capture of the electron by the Coulomb field of the projectile is not the dominant process at high velocity; simultaneous interactions with both the target and projectile Coulomb fields, the so-called second-order processes, are important.

In attempting to understand charge transfer it was discovered that electron capture is not confined to bound final states but that it extends into the ionization continuum of the projectile. This process, first observed in the early 1970s, gives rise to a singularity or cusp in the spectrum of the ionization electrons that is centered at an energy that corresponds to an electron moving at the projectile velocity. The continuum capture process accounts for a large fraction of the ionization events at low energies, yet it was completely overlooked in earlier treatments of ionization.

The most recent experimental evidence for the second-order nature of the electron capture is the observation of the Thomas peak. The projectile first scatters a nearly free electron, so that the electron attains the projectile speed, and then the projectile captures the electron with the help of the large Coulomb field of the target nucleus, which serves to scatter the electron parallel to the projectile trajectory. For proton projectiles the peak occurs in the angular distribution of the capturing projectile at an angle of 0.5 mrad (5 cm over the length of a football field).

Slow-Recoil Ion Production

As a fast, highly charged projectile passes through a neutral target of a light element, neon, for example, it can in a *single* collision remove nearly every target electron. The collision can transfer an enormous amount of energy to the electrons of the target, several kiloelectron volts, while transferring little translational energy to the nucleus of the target, a few electron volts or less. Thus, a fast heavy-ion beam is an efficient "hammer" for producing slow, highly excited ions. These ions are useful for spectroscopy and for the study of collisions with neutral targets.

Slow highly charged ions have been contained in electrostatic and electromagnetic traps for periods up to seconds. Metastable ions, such as the heliumlike neon ion, have been observed to capture an electron from a background gas and to radiate x rays and light. The captured electron goes into a highly excited orbit, thus forming a sort of population inversion in the final-state ions. The method has been used to create small external beams of slow highly charged ions, up to bare neon and heliumlike argon. Ultraviolet and x-ray emission from the slow ions do not suffer from the Doppler-shift problem of light from fast-ion beam sources.

Slow ions are just beginning to be exploited for precision spectroscopy. For example, Lyman-alpha radiation in argon has been observed. Precise measurements of its wavelength provide a measurement of the 1s Lamb shift in a new regime. In addition, low-recoil atoms may provide possible sources for new short-wavelength laser systems.

Tunable X Rays

Relativistic electrons and positrons can produce intense radiation as they pass through crystals along potential channels. To a laboratory observer these particles behave like very-high-frequency one- and two-dimensional oscillators. As the electrons move through the channels, they can weave about the planes, or revolve around the strings of positive lattice sites. An electron bound to a lattice plane forms a kind of one-dimensional atom, and an electron bound to an atomic string forms something like a two-dimensional atom. The electrons can emit tunable x rays, with characteristic energies up to 50 keV, highly directed in the forward direction. The x-ray spectrum reflects the electronic structure of the crystalline medium.

ATOMIC PHYSICS REQUIRING LARGER FACILITIES

Two areas require access to larger facilities than are usually employed in AMO research: accelerator-based atomic physics and AMO physics with synchrotron light sources. To provide the scientific background for the special recommendations on facilities for this research (Chapter 3), the new opportunities in these areas are summarized in this section.

Accelerator-Based Atomic Physics

Given the necessary tools, we will be able, in the next decade, to address an array of opportunities in the physics of highly charged ions. Advances in ion source and heavy-ion accelerator technology are placing at our disposal intense beams, both fast and slow, of ions of unprecedentedly high charge state. Hydrogenlike uranium and bare uranium ions have already been produced in relativistic heavy-ion beams, and slow beams of fully stripped argon from the new generation of ion sources have been produced. With fast relativistic beams, the structure of very-high-charged few-electron systems, for which QED and relativistic effects on level structures and on decay rates are huge, will be studied. Collision processes such as charge transfer with relativistic heavy-ion beams of unprecedentedly high charge will be

open to study. Beams of fast heavy ions, as well as synchrotron radiation, will be used to form ions in traps, producing highly charged but cool ions on which precision spectroscopy and collision experiments may be performed. Intense beams of both fast and slow multiply charged ions will enable merged-beam and crossed-beam studies of capture, ionization, and dielectronic recombination in collisions between multiply charged ions and between electrons and these ions. Such processes, while common in nature's hotter theaters, are only now coming into our grasp in the laboratory.

Slow beams from the new generation of multiply charged ion sources will make possible the study of the structure of multiply excited highly charged systems and of collision processes, both inner and outer shell, involving vacancies with binding energies up to tens of kiloelectron volts. (See Figure 4.5.) Such projectiles carry enormous amounts of electronic energy (14 keV for bare argon, about half a million electron volt for bare uranium) and will certainly give rise to new and violent physical phenomena when they encounter atomic or solid collision partners. New collision regimes will be opened: slow beams with inner-shell vacancies will allow us to probe for the first time collisions in which the inner-shell energy is shared with essentially all the colliding system's electrons during the collision, and energetic electrons and x rays emitted from the composite colliding system will be the rule rather than the exception.

Technological advances in data gathering, such as position-sensitive detectors and computer-based multiparameter data-acquisition systems, will bring us even closer to the ideal experiment in which all

---

FIGURE 4.5 Atomic Physics with Highly Charged Ions. Atomic collision processes involving multiply charged ions are important in astrophysical scenarios—particularly in stellar interiors—and in earthbound thermonuclear devices. These processes can be studied using accelerated ion beams. The upper figure shows the acceleration column of the Holifield Heavy Ion Research Facility, tandem Van de Graaff accelerator, which delivers beams of fast heavy ions with an energy per charge of approximately 25 MeV. The fast ions can be further stripped of electrons by passing them through gases or foils and then used by atomic physicists to probe ionization and capture collisions.   Recent advances in the design of ion sources now make it possible to produce intense beams of highly charged ions that, in contrast to ions from a high-energy source, move very slowly. These sources provide important new opportunities for investigating atomic-collision processes and to carry out spectroscopy on multiply charged species. The lower figure shows an electron cyclotron resonance source, one example of the new generation of low-energy ion sources that are just becoming available. [Photos courtesy of Oak Ridge National Laboratory (top) and the Lawrence Berkeley Radiation Laboratory (bottom).]

collision parameters before and after the encounter are experimentally determined.

Atomic, Molecular, and Optical Physics with Synchrotron Radiation

Synchrotron radiation now plays an indispensable role in atomic and molecular physics as an intense, reliable source of electromagnetic radiation spanning decades of the spectrum between the ultraviolet and hard x-ray ranges.

A new generation of synchrotron-radiation sources is technically feasible. Intensity can be increased by orders of magnitude, and the radiation can be delivered in reliable picosecond pulses. Circularly polarized as well as linearly polarized x rays can be produced. The way would be open for studying the role of many-body effects in atomic structure and dynamics, exploring the limitations of independent-particle self-consistent-field models, and testing modern theoretical approaches to electron-electron Coulomb correlation.

Synchrotron light beautifully complements the other major source of light used in photophysics—lasers. Laser sources are brighter and better resolved, but they operate only in a limited (though expanding) part of the spectrum. Synchrotron radiation is continuously tunable far beyond the limits of laser sources. Synchrotron-radiation sources can produce reliable picosecond pulses 10^9 times per second, again complementing the much lower pulse rate of the more intense laser radiation. An electron storage ring (in France) has recently been used successfully to operate a free-electron laser.

The scientific potential of this technology is great: photoexcitation of virtually any atomic or molecular subshell is possible, often with sufficient intensity to measure all secondary products (photons, electrons, and ions) including energies, ejection angles, and polarization or spin states. The incident wavelength can be freely tuned to excite resonances, threshold regions, or multiply excited final states. Many advances have already resulted from this powerful capability. For example, significant insights into correlations have emerged from studies of such diverse phenomena as autoionization, continuum-continuum coupling, postcollision interaction, and multielectron excitation. Molecular physics has advanced through studies of vibrational autoionization, shape resonances, and the breakdown of the single-particle model due to strong vibronic coupling for inner-valence ionization. The picosecond time structure of synchrotron light is only now beginning to be used in studies of intramolecular relaxation and energy-transfer processes.

The possibility of reaching the innermost shells of large atoms with hard synchrotron radiation provides access to processes in which relativistic and QED effects are prominent. Opportunities exist for measuring the frequency dependence of the Breit interaction and the screening of the self-energy in shells for which it has to date eluded calculation. Inner-shell threshold excitation with hard synchrotron radiation makes it possible to explore the domain where excitation and de-excitation of an atom occur in a single process, as in resonant x-ray and Auger Raman scattering. Finally, the joint use of lasers and synchrotron light has just been initiated, giving access to photoelectron spectroscopy of excited states of atoms in a manifold of hitherto inaccessible levels.

5

Molecular Physics

Molecular physicists apply the tools of physics to the problems of chemistry in order to obtain quantitative pictures of the structure of molecules, to learn how molecules interact and react, and to gain understanding of more complex states of matter such as liquids. Molecular and atomic physics share many experimental techniques, and their styles of research are similar in the depth of analysis and the desire for complete understanding of the phenomena in terms of the basic laws of physics.

Molecular physics straddles the border between physics and chemistry. In universities, the work is pursued in both physics and chemistry departments. In Europe the majority of the research is carried out in institutes of physics and physics departments; in the United States the research is most often carried out in chemistry departments. Research in the United States is funded by physics programs as well as by chemistry programs. We have attempted to portray here the activities in molecular physics that lie closest to the physics-chemistry interface.

THE NEW SPECTROSCOPY

Within the last decade laser spectroscopy has been combined with innovative molecular-beam techniques to create and study a multitude of new molecular species. For the first time, molecular physicists are able to prepare virtually any desired simple molecule in any desired quantum state. (See Figure 5.1.) Our understanding of molecular structure is advancing so rapidly that progress is not merely quantitative, it is qualitative. Traditional concepts of molecular structure are being challenged, in some cases set aside. An underlying unity between molecular structure and dynamics, long regarded as disparate areas, is beginning to emerge. The generation of archival data—transition frequencies, intensities, molecular constants, potential energy surfaces—continues its central role in spectroscopic research, but this, too, is being revolutionized by tunable lasers and modern data-processing methods. Altogether, these developments have stimulated an explosive growth in fields such as chemical kinetics, photophysics, and photochemistry and in a host of applications including combustion and plasma diagnostics, atmospheric monitoring, and laser development.

New Views of Electronic Structure

Laser spectroscopy has made it possible for the first time to examine systematically large classes of related molecular species, including species such as ions and radicals that can be produced only in trace quantities. By way of illustration, we shall briefly describe some new views of electronic structure that are emerging from the study of three of these novel species: Rydberg molecules (highly excited molecules), long-range molecules, and open-core molecules.

In a Rydberg molecule, one electron is in an orbit whose radius is much larger than the core molecular ion. The energy levels form a hydrogenlike pattern, but there are deviations from this pattern that can be measured with high precision. The deviations arise because the core is not a point charge: it has an extended nonspherical shape, and it is polarizable. Furthermore, the core vibrates and rotates. Frequently, these nuclear motions are fast compared with the orbital motion of the Rydberg electron. The Born-Oppenheimer approximation (nuclear motions slow compared with electronic motion), the basis of the traditional understanding of molecular structure in terms of potential energy surfaces, no longer applies. Fortunately, a highly successful theoretical framework for understanding Rydberg molecules has been created (multichannel quantum defect theory, mentioned in Chapter 4 in the section on Atomic Structure). The molecular-level structure can be viewed in terms of a slow electron repeatedly scattering from a molecular ion. The spectroscopy of Rydberg molecules, however, contains far more information, and yields a far clearer

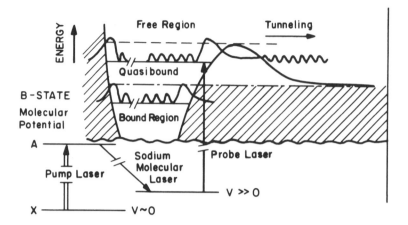

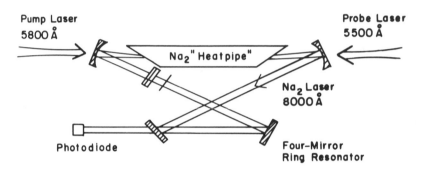

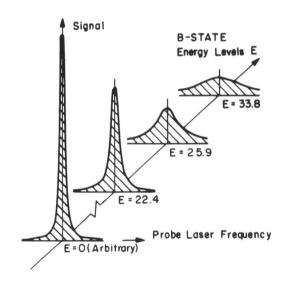

picture, than one could hope to attain from low-energy electron scattering by molecular ions.

Molecules in high vibrational states for which the nuclei spend most of their time far apart, typically five atomic diameters or more, have distinctive properties and form a class known as long-range molecules. Their properties can be understood from the properties of the isolated atoms, for instance the atomic polarizabilities and quadrupole moments. The attractive forces between atoms can be determined from the molecular-energy-level pattern of long-range molecules more precisely than by any other means, not only for ground states but also for excited states. This knowledge of the long-range forces is invaluable because it is these forces that control the rates of atom-atom recombination, transport coefficients in gases, and the cross sections for atom-atom inelastic collisions.

The third species, molecules in open-core states, is characterized by having one atom with a partially filled inner shell, for instance a transition-metal or rare-earth atom. The compact core of the open-shell atom can be regarded as an atomic ion that is perturbed by the rest of the molecule. The core, which is highly anisotropic because of its angular momentum, serves as a probe of its chemical environment, much as a nucleus in nuclear magnetic resonance. A myriad of low-lying electronic states can exist even in simple diatomic molecules. (Samarium monoxide, for instance, has over 1000 electronic states below 3 eV.) The study of open-core molecules provides a comprehensive picture for the electronic charge distribution in these states. The picture reveals the underlying simplicity of the atomic-ion-in-molecule electronic structure that lies concealed beneath an extremely complex energy level structure.

FIGURE 5.1   New Views of Molecules. Lasers and laser spectroscopy make it possible to study molecular states never before observable. The illustration shows one of many examples: a technique for studying molecules in states in which they are vibrating violently. The formation and breaking of bonds in chemical reaction generally proceeds through states like these.

The sodium dimer molecule, Na_2, is studied in this experiment using two tunable dye lasers, and the molecule itself forms a third laser. The first laser ("the pump") excites molecules in the "heatpipe" to an A state, one of the electronic states that might be studied by conventional spectroscopy. The excited molecules, which are in a laser resonator, form a laser (the Na_2 laser), which causes them to radiate into a state with high vibrational energy. The resulting long-range molecule is then studied with a probe laser that explores the region of the electronic B state, where the molecules are about to fall apart or dissociate—the quasi-bound region. (It is in this region that atoms in a gas form into molecules.) The rate at which the molecules dissociate is found from the width of the spectral lines measured by the probe laser. Different states have very different line widths, as shown in the bottom figure, revealing very different rates. The Na_2 laser is one by-product of the research on molecular structure; many new laser systems have been discovered in the course of such studies.

A natural connection links open-core molecules to the structure revealed by x rays in high-energy electron-molecular scattering, as described in Chapter 4 in the section on Atomic Dynamics. The same details of the electric field near the nucleus are sampled: in open-core molecules the probe is a natural vacancy; in the scattering experiments the probe is a short-lived inner-shell vacancy.

Electronic-Structure Theory: *Ab Initio* Calculations

Molecular-structure calculations have advanced so rapidly in recent years that it is now possible to calculate virtually any observable property of any small system. The accuracy can be high—for example, 10 parts per million for the ionization energy of H_2—though typically the computed properties are not so accurate as those obtained by experiment. Nevertheless, the computed properties are often sufficient for diagnostic purposes and to supplement empirical data. Moreover, in many cases they provide the theoretical background needed for interpretation of experiments. Calculated constants such as spin-orbit couplings or electric dipole moments are valuable for identifying the electron configuration of an electronic state; oscillator strengths can sometimes be calculated more accurately than they can be measured.

Hydrogen-Bonded Molecules

Midway between the strong chemical forces that create chemical bonds and the weak forces of van der Waals interactions there exists a class of forces that is responsible for molecular aggregates and polymers. The most important of these are hydrogen and ion-dipole bonding, for they govern the properties of substances vital to life ranging from water to DNA. The microscopic properties of the hydrogen bond—the structures of H-bonded aggregates, the energy levels, the bonding and dissociation energies—are at last beginning to be understood.

Most experimental work in the field of hydrogen bonding has focused on liquids and solids, for it is in these systems that H bonding is most important. Hydrogen bonds are very sensitive to the environment, however, and so liquid and solid systems are not well suited for studying them. To understand the bonds clearly, it is essential to study them in the gaseous state where the H-bonded complexes exist as free, unperturbed entities.

Traditional infrared spectroscopy of gaseous complexes reveals only broad unresolved bands, similar to the spectra obtained from liquids.

In the last 10 years, however, microwave spectroscopy and molecular-beam resonance spectroscopy have succeeded in revealing the structure of the ground states of H-bonded dimers and, in some cases, have also yielded information on dipole moments and dissociation energies. The richest lode of information, however, has come from infrared rotation-vibration spectroscopy. For example, the spectrum of the hydrogen fluoride dimer, which is held together by the hydrogen bond, displays all the important vibrational modes in great detail. As a result, a clear picture has been constructed of the potential barriers that separate its conformations.

Through high-resolution microwave and infrared spectroscopy of hydrogen-bonded complexes, accurate models of the pair potential for H bonding, and also for van der Waals bonding, can be constructed. This pair potential governs many of the phenomena of condensed-phase chemistry: bulk association, conformation and steric effects, solvation, solubility, and physisorption.

The ultimate impact of this work will extend to biology, for H bonding is vital to many biological processes. It plays a key role in the structure and formation of biomolecules and accounts for the activity of some toxic substances. The anesthetic activity of substituted fluorocarbons is believed to be due to H bonding. It has been discovered that biomolecules can exhibit semiconduction when doped with electron acceptors in which H bonding plays a role. Semiconduction is believed to be important in intercellular communication and in the control of cell proliferation. Calculations of all of these processes require accurate data on the structure of the H-bonded complexes. These data are becoming available.

Vibrational Structure of Polyatomic Molecules

The vibrational motions of a molecule are traditionally described in terms of normal modes—simple combinations of the nuclear displacement that execute simple harmonic motion at frequencies known as normal-mode frequencies. If the vibrational motion is sufficiently large, one expects motions to occur at multiples of the normal-mode frequencies known as overtones.

In the past, the normal-mode picture adequately explained the vibrational fundamentals and low overtones found in the 500-4000 cm^{-1} region of traditional infrared absorption spectroscopy. Higher overtones and combinations were generally too weak to detect, and there was no reason to question the normal-mode picture. Recently, highly sensitive laser techniques have been developed for recording pure vibrational spectra in the visible-wavelength region. Perhaps the greatest surprise from this work was the discovery that the normal-mode picture can break down completely at modest levels of excitation, typically 10,000 cm^{-1}.

One might expect that when vibrational amplitudes are so large that anharmonic effects are important, energy is transferred from a normal mode to the entire molecule, and the molecule simply heats up. However, it has been discovered that the energy can appear to be localized in a single band. For molecules containing one or more C—H, O—H, or N—H bonds, the vibrational spectrum near 20,000 cm^{-1} is dominated by highly localized vibrations in high-frequency, unusually anharmonic, bond-stretching motions. The normal mode structure is replaced by a local-mode structure, in which the vibration appears as a large-amplitude stretching motion of a single bond.

The observation of local-mode structures in the spectra of large polyatomic molecules has ignited excitement about the possibility of inserting energy into a specific bond. This has led to controversy about the possibility of inferring the rates for redistributing vibrational energy within a molecule from the widths of high-overtone spectral features. High-resolution spectroscopic studies have revealed that many of these features have sharp and assignable vibration-rotational fine structure, in contradiction to predictions based on classical mechanical calculations. In contrast to the situation for diatomic molecules, it is impossible to determine a potential energy surface for polyatomic molecules from the observed rotation-vibration levels, except near the equilibrium structure, where the normal mode approximation is useful. Several new semiclassical and variational schemes for obtaining more complete potential surfaces from spectral data have been proposed. The explosive growth of multiple-laser techniques for systematically obtaining high-quality, readily assignable spectral data for highly excited rotation-vibration levels is causing a complete rethinking of the problem of how polyatomic molecules vibrate.

MOLECULAR PHOTOIONIZATION AND ELECTRON-MOLECULE SCATTERING

Understanding the joint motion of electrons and nuclei in molecular fields is the essence of molecular physics. The dynamics of these motions underlie the spectroscopy, the physical transformations, and even the chemical changes in molecular systems. One strategy for studying these dynamics is to photoionize the molecule or to scatter electrons from it. Such experiments can provide physical insight into the processes occurring during molecular excitation and the escape of the electron through the anisotropic molecular field. The approach is straightforward in concept, but extracting a clear picture of the dynamics requires formidable experimental and theoretical tools. Major progress toward these goals during the last decade has yielded new understandings of electron motion in anisotropic molecular fields and the interchange of energy between electronic and nuclear modes. Moreover, recently developed techniques such as resonant multiphoton excitation portend accelerated progress in the future.

Molecular Photoionization

Photoionization is a powerful probe of the rotation-vibration-electronic dynamics of molecules. The photoelectrons, which are excited into well-defined optical ionization channels, carry to the detector information on the quantum state of the residual ion as well as on the dynamics of the photoionization process. Experimental developments including intense synchrotron light sources, pulsed dye lasers, and detectors of unprecedented sensitivity have led to rapid advances. It is now feasible to perform triply differential photoionization studies in which the wavelength, the photoelectron energy, and the ejection angle are independently varied. Previously measurements were only possible at fixed wavelengths from line sources; today the measurements can be carried out anywhere from the visible to the x-ray region. These experimental advances have been accompanied by the development of complementary theoretical methods.

Molecular photoionization studies are broad in scope. Here we discuss three topics of particular interest: autoionization, shape resonances, and resonant multiphoton ionization.

Molecular Autoionization Dynamics

Autoionization affects all molecular photoionization spectra, often producing dramatic spectral features. In the simplest case, autoionization occurs when a discrete state with positive total energy is coupled by a perturbation to the continuum of free-particle states. The perturbation allows the electron to escape. Autoionizing states usually consist of an excited Rydberg electron and an excited ion, which are bound together by their Coulomb attraction. Autoionization takes place during a close collision of the Rydberg electron with the ion: the excitation energy of the ion is transferred to the Rydberg electron, allowing it to escape from the ionic field. (Autoionization is the inverse of dielectronic recombination, a process discussed in Chapter 4 in the section on Atomic Dynamics.) A close encounter is essential, since only when the Rydberg electron is near the ion core can it participate fully in the dynamics of the ion and exchange energy efficiently with it.

A molecule can store the energy needed to ionize a Rydberg electron in any of its three modes—electronic, vibrational, or rotational. The most direct means of storing electronic energy is to create a hole in an inner molecular orbital, often by promoting an inner electron into a Rydberg state. Various combinations of vibrational and rotational excitation can accompany photoexcitation of Rydberg states; it is the existence and interplay among these alternative modes, or channels, that leads to the unique properties of molecular autoionization.

The most accurate and penetrating theoretical analysis of molecular autoionization has been carried out within the theoretical framework of multichannel quantum defect theory (MQDT). MQDT simultaneously treats the interactions between and within different excitation channels. The input to a MQDT calculation is a small set of physical parameters—quantum defects and dipole amplitudes—that characterize the short-range interactions between the excited electron and the core. From these few parameters, MQDT can yield values for many quantities that are directly related to the observables, for instance total photoionization cross sections, vibrational branching ratios, and photoelectron angular distributions. The complete elucidation of the autoionization spectrum of molecular hydrogen is one of the major triumphs of MQDT.

Shape Resonances in Molecular Fields

A shape resonance is a quasi-bound state in which a particle is temporarily trapped by a centrifugal potential barrier, that is, by the shape of the potential. In the case of molecular photoionization, the ejected photoelectron is partially blocked by a centrifugal barrier near the edge of the molecule so that, on the average, it must traverse the molecule several times before it eventually escapes the molecular core by quantum-mechanical tunneling through the barrier.

During the last 10 years shape resonances have become recognized as an important general class of phenomena in molecular physics, for a large variety of molecular properties have been found to be affected by them: x-ray and VUV absorption spectra, photoelectron branching ratios and angular distributions, non-Franck-Condon vibrational effects in molecular photoionization, elastic electron scattering, and vibrational excitation by electron impact, to name some of the most prominent.

Shape resonances provide a unifying link among different states of matter and among the various processes mentioned above. Because the resonances are localized in the strong potential of the molecular core they suffer only secondary effects owing to changes in the molecular environment. Hence, the same manifold of shape resonances in the photoionization cross sections of the free molecule is frequently observed during adsorption on a surface, condensation into a solid, or under completely different excitation conditions such as in electron scattering. To cite one example, there are four prominent shape resonances in the sulfur L-shell photoabsorption spectrum of gaseous SF_6. These are indistinguishable from the shape resonances in the spectrum of solid SF_6, and they also emerge in elastic electron-SF_6 scattering. This property of shape resonances has proven useful for studying the orientation and adsorption-site interactions of physisorbed molecules and for interpreting prominent features in spectra of ionic crystals.

Progress in understanding shape resonances has been profoundly influenced theoretically by the development of methods for treating molecular continuum states and experimentally by the harnessing of synchrotron radiation to study photoionization dynamics as a continuous function of wavelength, electron ejection angle, and electron kinetic energy. An excellent example in the strong interplay of theory and experiment is the prediction and confirmation of large changes in vibrational branching ratios and in photoelectron angular distributions induced by shape resonance. These arise because shape resonances are so sensitive to the internuclear separation that they behave differently in the individual vibrational ionization channels.

Shape resonances are a powerful probe of short-range excitation dynamics in molecules. They provide an important link among molecules in different physical states and among different physical processes. Our current knowledge is merely the tip of the iceberg. The expansion, refinement, and unification of these recent developments will provide an important theme in molecular physics in the coming years.

Resonant Multiphoton Ionization

In multiphoton ionization, a molecule absorbs several quanta of radiation to reach the ionization continuum. Intense, tunable radiation, which can be provided by modern dye lasers, is essential for inducing multiphoton ionization. Resonant multiphoton excitation via a single rotation-vibration level of an intermediate molecular state can dramatically simplify the spectrum, eliminating entire groups of rotational and vibrational states. Resonant multiphoton ionization reveals the photoionization dynamics of excited states that are fully specified quantum mechanically. Photoionization branching ratios, photoelectron angular distributions, alignment, and fragmentation—all the important properties in single-photon ionization—can now be studied by photoionization of excited molecular states. Resonant multiphoton ionization provides a means to probe dipole-forbidden ionization channels, including whole manifolds of autoionizing states that have never been observed. Collisional effects on the resonant intermediate state can be studied, for instance by observing changes in the photoelectron angular distributions. Because the total energy of the several photons can readily exceed the energy of a single one, states of much greater energy can also be probed. There are many other applications, such as continuum-continuum transitions and nonlinear and nonresonant effects, but these few examples should suggest the great potential of this emerging stream of research.

Electron-Molecule Collisions

The electron-molecule continuum is highly structured, containing transient negative-molecular-ion states that reveal themselves by a rich pattern of resonances in electron-molecule scattering measurements. Electron-molecule resonances are in many ways analogous to electron-atom resonances, but the molecular resonances possess additional structure because of the underlying nuclear dynamics. Moreover, the electron-molecule interaction is inherently anisotropic. It is anisotropic not only close to the molecule where the strong screened nuclear attractive potential dominates but also at large electron-molecule separations where the permanent dipole, quadrupole, and higher moments of the molecule give rise to the long-range terms in the interaction potential. Even the weak polarization interaction that characterizes the adiabatic response of the bound molecular electrons to an approaching electron is anisotropic.

During an electron-molecule collision the nuclei are free to move, though the time scale of this motion is much longer than that typical of bound electrons. The fact that nuclei are separated from one another by distances of the order of atomic dimensions essentially guarantees that the average electron-molecule interaction is strong over a fairly large region of space. This frequently gives rise to the potential resonances that were described in Chapter 4 in the section on Atomic Dynamics. The interplay of the electron transit, the lifetime of a resonance, and the various nuclear response times presents a theoretical challenge that is just beginning to be met.

The most important advances in our understanding of electron-molecule collision phenomena are in those aspects of the target that are unique to molecules. For electron energies near a resonance, the appropriate electronic wave function for small electron-molecule separations is that of the transient negative ion; the nuclei move in a distorted potential field. As the electron leaves the complex, the molecule is left in a coherent superposition of states. It has been found from the measurements and from the study of semiclassical models that the resonant structure due to nuclear motion is extremely sensitive to details in the theoretical description. This sensitivity has provided both a clear picture of the dynamics and a precise test of the *ab initio* description. The experiments have produced surprises. One example is the discovery of a single, sharp resonance peak in vibrational-excitation cross sections for several polar and nonpolar molecules just above the excitation thresholds. Several models have been suggested, invoking virtual states and other mechanisms, but the issue remains unsettled.

MOLECULAR DYNAMICS

Chemical reactions involve complex many-body interactions. Because the quantity of information required to characterize all the particles in a reaction is enormous, the practical and intellectual goals of obtaining a clear picture appear formidable. By combining supersonic molecular beams with laser schemes for detecting the reaction products, however, the problem can become tractable. For instance, state-specific reactive scattering studies are now capable of distinguishing complicated reactive interactions according to the classes of forces and the stages of temporal evolution. It is possible to distinguish the signatures of energy being released while the reagents approach and while the reactive products separate. Similarly, it is now recognized that the branching ratios for different reaction products can depend critically on the orientations of the electron orbitals. In many cases, highly detailed studies of complex molecular phenomena have led to simple physical explanations and to new points of view.

State-to-State Chemistry

Chemical reactions are usually described by listing the initial and final products, though this actually gives little insight into how the reaction occurs. The complete enumeration of all the quantum numbers of the system, initial and final, would provide a far more thorough description, one that could allow a rigorous confrontation of theory with experiment. This has now been accomplished for the most elementary type of molecular encounter: collisions in which energy is transferred to a molecule from an atom or another molecule. If all the important quantum numbers are measured these are called state-to-state collisions. The study of state-to-state collisions in the past decade constitutes an important advance in basic molecular physics. The results have already provided new insights into molecular dynamics, and they are expected to be valuable in applications involving energy transfer in gases.

Experiments on state-to-state collisions have been made possible by two advances. The first is high-intensity supersonic molecular beams. These beams have an extremely narrow velocity distribution, and they can achieve very low internal temperatures—rotational temperatures of a few kelvins are typical. (See Figure 5.2.) As a result, the molecular states have a much smaller thermal spread in quantum numbers than otherwise possible. The second advance is the tunable dye laser. These lasers make it possible to resolve completely the quantum states of the collision products. The lasers can also serve as precise velocity analyzers by utilizing the Doppler shift, and they can be used to prepare the system in high vibrational states by optical pumping.

State-to-state collisions of HD with He provide one example of the power of the technique. Differential cross sections for transitions from individual initial rotational state to each final state have now been measured with high resolution. Furthermore, the cross sections have been calculated using a full quantum-dynamical formulation and a potential surface generated from first principles. Comparison of experiment and theory pointed to the need for some corrections to the potential surface, but with these, theory and experiment agreed even to minute details of the quantum diffractionlike oscillations in the data. Similar studies have been carried out with other simple systems.

Rotational state-to-state collisions represent the most elementary form of energy transfer in molecular systems. Such energy-transfer processes play critical roles in the dynamics of supersonic expansions, gas lasers, atmospheric physics, and planetary atmospheres and in

FIGURE 5.2 Molecular-Beam Scattering. Molecular beams, originally created to study the proper ties of isolated atoms and molecules, are now extensively used to study interactions between molecular species and the dynamics of chemical reactions. This molecular-beam scattering apparatus is designed for studying collisions in systems such as helium and molecular nitrogen. It uses supersonic atomic or molecular beams to provide intense streams of neutral particles that have only a small spread in their speeds. Particles in two separate beams collide under highly controlled conditions, and the speed and direction of the scattered particles are measured with high resolution. The experiments provide detailed information on how energy is transferred between atoms and molecules, for instance, how much of the energy goes into translational motion and how much goes into rotation. Experiments such as these guide the development of the theory of energy transfer and provide an important step toward understanding the precise steps that occur in a chemical reaction. The information is also useful for understanding the drag on airplanes and spacecraft. Early molecular-beam experiments were of the table-top variety, but, as the picture indicates, the apparatus now can be large and elaborate. (Courtesy of the Max-Planck-Institute for Fluid Dynamics, Göttingen, Federal Republic of Germany.)

many other problems. Enormous numbers of cross sections may be required to understand a typical problem. State-to-state experiments frequently generate huge arrays of data—100 cross sections for a single system are not unusual—and the sheer volume can be overwhelming, obscuring the confrontation with theory and complicating attempts to model molecular energy transfer in specific situations. Fortunately, in the course of studying state-to-state collisions in systems like Na_2-Xe and LiH-He, a universal type of behavior was discovered. By combining ideas of angular momentum addition and topology, a general explanation for this behavior has been found. Out of this has emerged a model that provides simple classification of the data, realistic extrapolation and interpolation of the measurements, and accurate analytical forms. One of the surprises from state-to-state research is that the empirical rules previously used to estimate energy transfer rates were misleading and could have led to potentially serious errors in applications.

Radiative Collisions

Beginning in 1972, a series of papers in a Soviet journal suggested that intense radiation could affect inelastic atom-atom collisions. The effect was demonstrated 5 years later in the United States in a study of collisions between excited strontium and ground-state calcium in the presence of intense radiation from a dye laser. Without the laser light the collisions were elastic; with it, a large energy exchange occurred: the excitation was transferred from the calcium to the strontium. The surprise is that the laser light was resonant with neither strontium nor calcium atomic transitions. The light was resonant with molecular levels of the strontium-calcium system as they evolved during the collision. The experiment can be viewed as the spectroscopy of a chemical reaction in progress—an event never before witnessed. It is now recognized that radiative collisions bear on large classes of basic molecular phenomena and that they also have the potential for useful applications in chemical processing.

Radiative collisions provide a new and flexible probe for studying a chemical reaction in progress. One can describe the collision in terms of the radiative excitation of an atom whose energy levels are tuned into resonance with the laser light by the changing perturbation of a second atomic species. Alternatively, one can describe the two colliding atoms as a single molecular entity whose energy levels come into resonance with the laser radiation.

Radiative collision can occur even if the radiation is not in resonance between two discrete atomic or molecular levels; the radiation can be in resonance with continuum levels. Radiative electron-atom scattering and collisional ionization are two examples. In the first, the energy spectra of electrons scattered off argon atoms in the presence of a pulsed carbon dioxide laser exhibit peaks corresponding to both absorption and stimulated emission. Electrons that have changed their kinetic energy by as many as 11 carbon dioxide photons have been detected.

Chemical reactions can occur during a radiative collision. Xenon and molecular chlorine have been observed to react to form XeCl in the presence of laser light, though no reaction occurs otherwise. Thus, radiative collisions have the potential of leading to new forms of photochemistry. In particular, one can envisage triggering the release of a great deal of stored chemical energy with a relatively weak light pulse. The process can occur very rapidly, perhaps rapidly enough to be useful in a very-short-wavelength laser.

New Ways to Understand the Dynamics of Chemical Reactions

Chemical reactions usually generate products in states of internal excitation; the products then give off heat, emit radiation, or go on to react further. The relative rates at which these internal product states are formed are important for applications ranging from research into new chemical species to industrial processes and the creation of chemical lasers. In the last decade, much has been learned about the dynamics of chemical reactions in the gas phase. (See Figure 5.3.)

The dynamics of a chemical reaction is controlled by the potential energy surfaces that describe the interaction between the reacting particles. These interactions determine the motion, whether or not the reaction occurs, and the detailed path from the initial states of the reactants to the final states of the products. In many cases the potential energy surfaces can be obtained by the techniques of modern quantum chemistry. Once the surfaces have been calculated, the rates of reaction and the detailed dynamics must be predicted. A variety of new theoretical approaches are available, almost all relying heavily on computers. We describe here three of these. In order of increasing difficulty, and increasing detail and accuracy, they are variational transition-state theory, quasi-classical trajectory calculations, and approximate quantum scattering calculations.

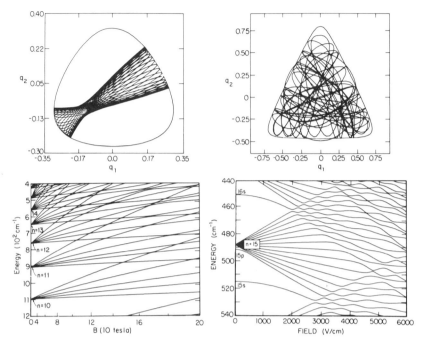

FIGURE 5.3 Classical and Quantum Chaos. The transition between orderly and chaotic motion is important in wide classes of classical systems. Similar behavior is beginning to be recognized in quantum systems. The upper left drawing is a plot of position against speed for a simple nonlinear system; the motion is orderly. If the system is slightly perturbed, however, the motion abruptly becomes disorderly, as shown in the upper right-hand drawing. Analogous behavior appears to be displayed by atomic systems. The lower left drawing shows energy levels of a highly excited hydrogen atom in a strong magnetic field. The energy levels appear to evolve in a simple and orderly fashion. The lower right-hand drawing shows energy levels for a highly excited sodium atom—essentially a perturbed hydrogen atom—in an electric field. The levels look orderly at low field, but as the field is increased they abruptly become disorderly. Order-disorder transitions appear to play important roles in molecular systems—for instance, the localization of energy of a highly excited polyatomic molecule in a single vibrational mode—and in optical systems such as optically bistable devices. (Courtesy of the Joint Institute for Laboratory Astrophysics.)

Variational Transition-State Theory

The transition-state theory of chemical reactions has been in existence for many years, but in the last decade the formal basis of the theory has been re-examined and the theory has been reformulated to yield much more accurate results. The variational transition-state theory requires finding the set of atom configurations that most effectively divides reactants from products. Using classical mechanics, the reaction rate is calculated by determining the rate at which systems can cross this dividing surface.

Quasi-classical Trajectory Calculations

The method of quasi-classical trajectories employs computers to trace the dynamics of collisions between a large set of reactants. The calculations determine which states will react and what the distribution of energy will be in the products. This approach, which provides an excellent semiquantitative guide, is now widely used. The method has recently been extended to electronically nonadiabatic collisions, collisions in laser fields, and a number of other processes. This approach will undoubtedly be used extensively in the next decade to predict reactivity and selectivity of relatively simple chemical reactions.

Approximate Quantum-Scattering Calculations

The dynamics of chemically reacting systems can, in principle, be predicted precisely by using accurate potential energy surfaces and quantum scattering theory. The calculations are extraordinarily difficult for larger systems, but a number of approximations have been developed that greatly simplify the problem yet preserve the accuracy. "Sudden approximations" permit mapping a set of simple collinear quantum reactive scattering calculations onto three-dimensional space to yield approximations to the true scattering cross sections. Collinear quantum calculations have been used to correct results from classical transition-state theories and to investigate the role of resonances in reactive scattering.

Recently the approximate quantum calculations revealed a new type of molecular state, a true bound state of some triatomic systems but with binding occurring only because of the vibrational motion. The molecule FHF appears to be a prime candidate for exhibiting such states. The study of this phenomenon, and of other quantum effects in reactive scattering, will undoubtedly lead to better control of chemical reactions and, one hopes, the eventual development of systems in which the various state-to-state reactions can be selectively activated.

Resonances in a Simple Reaction Complex

Some time ago chemical physicists developed techniques for calculating cross sections for simple chemical reactions such as $H_2 + F \rightarrow HF + H$. The results contained an unexpected conclusion: for systems constrained to move along a line the reaction probability changes abruptly as the initial translational energy is varied. It was discovered that this behavior is due to the formation, at certain energies, of long-lived dynamic resonances in which the colliding reagents remain close together for several vibrational periods. The resonances arise from a temporary trapping of the energy of the system in the internal degrees of freedom. Such resonances occur elsewhere in physics, in nuclear reactions, and in electron-molecule collisions, for instance, but they had never been observed for chemical reactions.

The resonances were discovered in recent experiments using crossed molecular beams for the reaction $H_2 + F \rightarrow HF + H$. The theoretical studies reveal that the resonances are sensitive probes of the potential energy surfaces in the region where the atoms are close to each other. The agreement between theory and experiment illustrates the increasing predictive power of dynamical chemical theory.

Bond Breaking and "Half-Collisions"

The breaking of a molecular bond is an essential step for most chemical reactions, but until recently the process could not be witnessed directly. Now, however, the structure and molecular motion during dissociation can be observed. Because the latter half of many chemical reactions involves a dissociative state whose products fly apart, the photofragmentation process has come to be known as a "half-collision." Laser light is used to break selectively the bonds in simple molecules. The speed and angular distribution of the molecular fragments can be measured directly or found indirectly from the spectrum of the dissociating molecule. From this it is possible to determine details of the dissociative state and whether dissociation occurs quickly or slowly. Molecular-bond-breaking processes as short as 10^{-14} s have been studied.

Not only is the laser useful as a scalpel to break bonds, it is a versatile tool for interrogating the fragments of the bond-breaking process. Detailed information on the states of the fragments can be obtained by re-exciting the molecules or atoms and observing fluorescence from particular excited states. Because the laser light is naturally polarized, specific orientations of the molecular fragments can be obtained as well. Thus it is possible to "photograph" the structure of the highly excited molecule as it is on the verge of breaking up, to measure the forces experienced by the fragments as the bond is broken, and to study the photofragment states that result. This is invaluable for understanding, and possibly for influencing, the way that reactions distribute the energy released among the product molecule electronic, vibrational, rotational, and translational motions.

In addition to their contribution to our basic understanding of chemical reactions, molecular photofragmentation techniques have played key roles in applications such as the study of photochemical smog formation and the development of new lasers. Photodissociation measurements provide much needed information on the kinds and concentrations of reactive radical species that are created in the atmosphere by sunlight and the reactions that form harmful pollutants or deteriorate the ozone layer. Molecular photofragmentation lasers were one of the first of the chemical lasers. The art has advanced to the point that it is now possible to activate molecular gas lasers by direct irradiation with sunlight. Solar-pumped lasers may become practical for space-based communications.

Reactions at Very Low Temperatures

Traditional thinking about ion-molecule reaction at very low temperatures suggests that their behavior should be simple and predictable: either the rate coefficient at temperatures below 300 K should be as predicted by simple orbiting theory or there must be an energy barrier that makes the rate completely negligible at temperatures of less than about 100 K. Recent studies indicate that surprises are in store. One new technique utilizes Penning traps in which the ions can be cooled to below 10 K and held for many hours while reacting with cold gas at low densities. A study of $NH_3^+ + H_2 \rightarrow NH_4^+ + H$ at 300 K and above indicated a substantial energy barrier for this process. It had been assumed that this process was totally negligible at the 10-20 K temperatures of dense clouds in the interstellar medium. However, the trapped-ion studies at temperatures of 10-20 K, coupled with other studies at 80-300 K, have shown that as the temperature is decreased the rate coefficient goes through a minimum and then starts *rising* very steeply. This process is important at the temperatures of the interstellar medium. It is believed that a complex is formed at low temperatures that lives long enough to permit quantum-mechanical tunneling through the centrifugal barrier.

SOME NOVEL MOLECULAR SPECIES

A large variety of novel molecular species and molecules in unusual classes of states have emerged from the laboratories of molecular physicists within the past few years. In addition to the species discussed in the first section of this chapter, the list includes positive and negative ions, neutral and ionic clusters, metal atom dimers, van der Waals molecules, free radicals and other highly reactive species, unstable isomers, metastable molecules, and polyatomic molecules in selected highly excited rotation-vibration levels. From this list we chose two for discussion: molecular ions and van der Waals molecules.

Molecular Ions

Molecular ions play key roles in the chemistry of solutions, in atmospheric chemistry, in the interstellar medium, in plasmas, and in flames. Until recently the experimental study of ions in the gas phase was extremely difficult because ions could not be prepared at high concentration or in isolation from other molecular species. The only practical way to study molecular ions was in the solid state—in crystals. Now, however, the spectroscopy of free positive ions is flourishing. A technique called laser magnetic resonance has opened the way to the study of a host of ions and free radicals, including a number that are of astronomical interest. Another method employs laser spectroscopy in an ac discharge. Owing to the Doppler effect, the alternating electric field shifts the transition frequency of the ions in and out of the narrow laser resonance, allowing the signal to be separated from the intense background that is due to the abundant molecules. Virtually any stable neutral particle with one extra proton can be created, and the method works with numerous other molecular ions and radicals. A third method employs a pulsed supersonic molecular beam. The ions are formed in a plasma and then cooled and isolated in the beam's expansion region. As the supersonic beam expands, the temperature drops rapidly, cooling the rotational and vibrational modes. This reduces the number of states occupied, vastly simplifying the spectrum. Rotational temperatures as low as a few degrees are routinely achieved. The cold gas can condense into well-defined cluster-ion species. It is actually possible to watch how a vibrational frequency of a free molecular ion evolves as the ion joins increasingly large clusters of inert gas atoms, finally reaching the limit where it is essentially isolated in a rare-gas matrix. (Clusters are discussed further in Chapter 7 in the sections on Condensed-Matter Physics and Materials Science and on Surface Science.)

In addition to optical spectroscopy, positive molecular ions have recently been studied by other approaches. For example, ultraprecise infrared spectroscopy of the elementary ions H_2^+ and HD^+ has been carried out using light from an accurately known fixed-frequency laser. The light is tuned into resonance by varying the speed of the ion beam.

Van der Waals Molecules

Within the past decade the study of weakly bound molecules has emerged as a new tool for understanding the principles of chemical structure. These molecules are composed of stable molecules, or inert atoms, which are held together not by the covalent or ionic forces that normally hold molecules together, nor by hydrogen bonding, but by the weak van der Waals force.

The list of van der Waals molecules that have been studied include species such as argon attached to a variety of atoms and tightly bound molecules, dimers such as hydrogen, $(H_2)_2$, or hydrofluoric acid, $(HF)_2$, and such unlikely chemicals as HF-ClF. In fact, essentially any simple combination of atoms and molecules can now be studied.

The structure of a van der Waals molecule is often unexpected. For instance, one would expect argon to attach itself to the middle of ClF simply because that would put it closest to most neighbors, forming a T-shaped molecule. It does not—the molecule is linear, with the argon attached to the chlorine. Because hydrogen bonding is relatively strong, one would expect HF to bond to FCl with hydrogen shared between two partners, for instance FH-FCl. It does not—the structure is HF-FCl. The benzene dimer, by contrast, is much simpler than one might expect. The planes of the ring are perpendicular, forming a T. The conformation is the same as the crystalline solid.

Van der Waals molecules provide new opportunities to study how molecular pairs interact and the configurations that they assume. The significance of the work lies in this: Chemical structure remains a fundamentally unsolved problem. There is no way to predict the geometric conformation of molecules from general principles, and there is no perturbation theory for chemical bonding—every species behaves like a new system. By providing an opportunity to test approximate theories on a large class of relatively simple systems, van der Waals molecules provide an advance toward understanding molecular structure in all of its manifestations, including the liquid and solid states, and toward understanding chemical reactions.

6

Optical Physics

Optical physics encompasses the physics of electromagnetic radiation and the interaction of matter and light. It includes the generation and detection of light, linear and nonlinear optical processes, and spectroscopy. The distinction between optical physics, applied physics, and optical engineering is blurred, for devices and applications are close companions to basic research in this area of physics.

The first two sections of this chapter deal with lasers and laser spectroscopy—topics that have transformed optical science. The last two sections are devoted to quantum optics and coherence, and to femtosecond optics. Nonlinear optics, a major area of modern optics, encompasses so many different streams of research and applications that we have not attempted to describe it separately. Nonlinear optics plays a role in most of the topics discussed in this chapter.

LASERS—THE REVOLUTION CONTINUES

From checkout scanners at supermarkets to laser disk recordings, lasers have become commonplace, but the scientific revolution they precipitated is continuing, propelled not only by the discovery of more and more applications but by the steady development of new lasers and new laser techniques.

The development of tunable lasers that can operate throughout the visible and into the infrared and ultraviolet ranges has had a major

110

impact on basic science during the past decade; instances of the advances are scattered throughout this report. Dye lasers are the most ubiquitous of these tunable light sources. Continuous-wave dye lasers achieve a stability and resolution far exceeding those of traditional light sources—improvements by factors of hundreds to hundreds of thousands are typical. (See Figure 6.1.) Pulsed dye lasers provide such intense radiation that nonlinear processes such as frequency doubling and multiphoton absorption are now widely employed. It is possible to use several lasers in one experiment, providing innumerable new strategies for studying atomic and molecular phenomena.

Many new laser sources have come into use during the past decade, from color-center and semiconductor lasers in the infrared to excimer lasers in the ultraviolet. The semiconductor diode laser is already a key component of a major new industry—fiber-optic communications—as discussed in Chapter 8. Notwithstanding these advances, other sources are urgently needed. We possess no efficient optical lasers, and many wavelength regions outside of the visible are difficult to achieve or are inaccessible. There is wide interest in ultraviolet lasers, and the x-ray region continues to be a tantalizing goal.

Major advances in lasers have come from research in atomic, molecular, and optical (AMO) physics, and the level of activity and excitement continues to be high. Current developments include superstable optical lasers, hollow-cathode lasers that operate in the ultraviolet using sputtered metal ions, and pulsed-gas ultraviolet lasers. One of the most dramatic developments in laser technologies during the past decade has been the construction of gigantic neodymium glass lasers powerful enough to ignite thermonuclear fusion reactions. These devices stand as triumphs of optical engineering: they have achieved energy densities far greater than anything previously produced by man.

One final class of lasers must be mentioned—the free-electron laser. First demonstrated as an infrared laser, these devices are now being engineered for wavelengths from the far infrared to the vacuum ultraviolet. Free-electron lasers generate coherent radiation by stimulated emission from relativistic electrons traveling through a periodically varying magnetic field. They are attractive because their wavelength can be varied simply by changing the energy of the electrons and because high power and high efficiency appear to be possible.

Synchrotron radiation provides an alternative to laser light as a source for ultraviolet radiation. Because of their high brightness at short-ultraviolet and x-ray wavelengths, these sources are being used increasingly, particularly in condensed-matter physics and surface science.

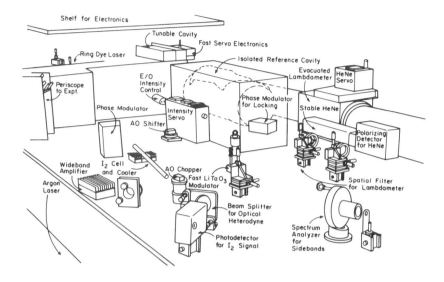

FIGURE 6.1 Superstable Tunable Lasers. This tableful of optical components is a tunable-dye-laser system that is so stable that the frequency of the light can be adjusted much like the signal of a radio-frequency or microwave generator. The jitter in the frequency of the laser is only 100 Hz. (The frequency of light is about 5×10^{14} Hz.) Highly stabilized lasers can be used to create and study new types of atomic and molecular species and to carry out ultrahigh-precision spectroscopy. They are also useful for applications such as stopping atoms and studying relativity and for gravity-wave detection.

Although this highly stabilized system is at the benchtop stage in a research laboratory, industry has been very effective in making advanced laser and optical technologies available rapidly. (Courtesy of the Joint Institute for Laboratory Astrophysics.)

This summary is by no means complete—one could mention numerous new solid-state and gas lasers, advances in the design of optical resonators, and in new pumping methods—but it should give some idea of the level of activity and the rapid progress in laser design.

Laser-based ultraviolet and x-ray sources are also being developed. For instance, intense, highly monochromatic tunable vacuum ultraviolet radiation has recently been generated by scattering laser light from metastable excited atoms. Another method uses an infrared pulse of a few Joules of energy from a neodymium-glass laser. When the light is focused on a heavy-metal target it creates a highly ionized plasma that emits a substantial fraction of the incoming laser energy as an intense short burst of soft x-ray radiation, emerging from a pointlike origin and covering a continuous spectrum. The method is disarmingly simple. (See Figure 1.3.)

Every one of these lasers and light sources has an interesting scientific and technological history. To illustrate the role of AMO physics in the development of lasers, we have chosen one device—the excimer laser—to describe in some detail.

Excimers and Excimer Lasers

Excimers are diatomic molecular systems for which the electronically excited state is tightly bound but the ground state is a very loosely bound, essentially unbound, van der Waals molecule. The emission spectrum for excimers is characteristic of transitions from bound molecules to free atoms; such molecular transitions are ideal for high-power gas lasers. Most excimer systems involve a rare-gas halide molecule (a molecule composed of a rare-gas atom and a halide atom, for instance, xenon-fluoride). The application of rare-gas halide excimer molecules to efficient, high-power lasers is a success story for high technology that has its roots in fundamental AMO physics.

Rare-gas halide lasers employ an electron beam or electrical discharge to deposit energy into rare-gas mixtures with a halogen-containing fuel. The electronically excited state of the rare-gas halide molecules is formed efficiently owing to the unique properties of rare-gas atoms and molecules. The stored energy in the rare-gas halide excimer molecules is then extracted by laser action.

Within a 5-year period following the discovery of the rare-gas halide emission spectra, small commercial lasers were available for laboratory use and large devices were under construction for military and national energy-related goals. This rapid development required wide collaboration within the AMO community, including specialists in the interactions of electrons, ions, ground and excited state atoms and molecules; the optical properties of the laser medium; and the hardware associated with electrical deposition of energy into high pressures of rare gases. The rapid development of rare-gas excimer lasers illustrates the value of maintaining a reservoir of trained personnel in AMO physics.

The development of rare-gas halide excimer lasers provides an excellent example of a situation where detailed kinetic data at the state-to-state level were vital. Such information will also be vital to the design of other energy-storage gaseous systems. Thus there is an urgent need for the knowledge needed to develop kinetic models at the state-to-state level for reactive systems involving atoms and small molecules.

A final point about molecular excimers is their role in stimulating theoretical work in bound-free emission spectroscopy. Models have been developed that permit bound-free spectra to be accurately simulated for several rare-gas halide excimers. Further advances would be of value not only because of the intrinsic interest of these molecules but because of the possibility of discovering new excimer systems.

LASER SPECTROSCOPY

Much of what we know about the structure of matter comes from spectroscopy. During the past decade both the techniques and uses of spectroscopy have advanced so rapidly that the term has acquired new meaning. Spectroscopy has truly undergone a revolution.

New spectroscopic techniques have achieved a precision and sensitivity enormously greater than the classical techniques of absorption and fluorescence; they have opened new areas in atomic and molecular physics. Unusual species such as Rydberg atoms and molecules can be created routinely, and familiar species can be viewed from new perspectives. For example, the ability to excite a single electronic molecular state with known quantum numbers has had a large impact on molecular physics, as described in Chapter 5 in the section on The New Spectroscopy. Beyond this, a new optical technology has emerged, combining atomic, molecular, and optical science and leading to innovations such as optical frequency standards, new light conjugators, four-wave mixers, and far-infrared detectors. The collective enterprise has come to be called laser spectroscopy. This term, however, is something of a misnomer, for laser spectroscopy extends far beyond the conventional idea of spectroscopy.

Ultraprecise Laser Spectroscopy

A unique feature of laser light is its spectral purity. Conventional monochromatic light sources typically achieve a spread in frequencies of 1 part in 10^5. Commercial dye lasers now routely achieve 1 part in 10^8. In advanced laboratories, dye lasers have been operated with a stability and spectral purity greater than 1 part in 10^{12}.

The art of wavelength measurement has also made impressive advances. For instance, automated digital wavemeters make it possible to determine wavelengths to 1 part in 10^7 or 10^8 in a split second. Photodiodes have been developed that can observe beats in the signals of two different lasers at frequencies as high as several terahertz (1 terahertz = 10^{12} cycles per second). As a result, lasers operating at quite different frequencies can be compared with high precision. In fact, the possibility of directly measuring optical frequencies in terms of the cesium microwave frequency standard has recently been demonstrated. Essentially, this creates a new optical technology in which the frequency of light is directly measured, much as is done with radio-frequency and microwave signals.

Ultrasensitive Spectroscopy

Lasers make it possible to observe extraordinarily weak absorption of light using a variety of simple techniques. For instance, by placing a small sample of a gas such as ordinary air inside the resonator of a dye laser, strong yellow absorption bands of water vapor and molecular oxygen appear in the laser's light. These bands are barely perceptible by classical techniques, even with absorption paths of many miles.

Photoacoustic detection is an ultrasensitive technique for observing the absorption of light in gases, liquids, or solids. A laser beam is modulated at an audiofrequency, and a microphone detects sound waves generated by the periodic small heating of the sample. Another ultrasensitive method is optogalvanic spectroscopy. Modulated laser light enters a gas discharge; when the laser frequency is tuned to a resonance between two energy levels, the discharge current or voltage displays a modulated signal, even when both levels correspond to excited states. The technique requires only a discharge tube, a laser, and an oscilloscope. Among its many applications, optogalvanic spectroscopy permits studies of sputtered metal atoms and transient species, such as ions and molecular radicals.

The ultimate in sensitivity can be reached with a related technique: resonant photoionization by intense laser light. A single gaseous atom of almost any element or isotope can, in principle, be selectively excited and detected, even in the presence of large numbers of atoms of different species. Potential applications range from trace analysis and the detection of impurities in semiconductor materials to the search for rare unstable isotopes. (The presence of these isotopes in mineral deposits has been proposed as a telltale indicator for solar neutrino reactions or of the double beta-decay process that would occur if the neutrino had a rest mass.)

Doppler-Free Laser Spectroscopy

In the past, spectral resolution in a gas was limited by the Doppler effect, the frequency broadening due to the motion of the atoms. Laser spectroscopy provides several methods for eliminating the first-order Doppler broadening, permitting observation of the much narrower natural width of the spectral line. Some of these methods are relatively simple, well suited to observing rare or short-lived species. Others have played roles in stabilizing the frequency of lasers to atomic or molecular transitions.

One method, saturation spectroscopy, employs a strong laser beam to label a group of atoms and a counterpropagating beam to pick up a signal from those atoms that have no Doppler shift. Saturation spectroscopy has been applied to problems ranging from the finest details of molecular structure to collisional effects and precise measurements of fundamental wavelengths in atomic hydrogen. Another method of Doppler-free spectroscopy employs two counterpropagating laser beams to induce a two-photon transition: the first-order Doppler shift essentially disappears for all the atoms or molecules. Using this method, the "forbidden" 1S-2S transition in hydrogen has been observed. This measurement represents an important spectroscopic advance because its intrinsic linewidth is more than a million times narrower than for normal optical transitions. It provides the opportunity to measure the Lamb shift in the ground state, and the wavelength can be directly related to the Rydberg constant and the electron-proton mass ratio. The 1S-2S transition in positronium has also been measured by two-photon spectroscopy. As discussed in Chapter 4 in the section on Elementary Atomic Physics, laser spectroscopy of positronium opens a most useful new line of research in the physics of elementary systems.

Laser Cooling

Although Doppler-free methods remove the first-order Doppler shift, the second-order Doppler shift remains. The effect is small—the frequency is typically shifted by 1 part in 10^{11}—but it can result in serious errors in high-precision measurements. The second-order Doppler shift is proportional to the kinetic energy of the particles, and the only way to reduce it is to reduce the particles' motion. This has been accomplished by using the momentum of laser light in various ingenious experiments to "cool" atoms or ions, that is, to slow them or even bring them to rest. Ions held in an electromagnetic trap have been cooled to the millikelvin range by absorbing laser light that is tuned slightly below resonance. (See Figure 6.2.) Recently an atomic beam of sodium was cooled, actually brought to rest, by laser light.

A prime motivation for laser cooling is to create better frequency standards, either at microwave frequencies or at optical frequencies. Optical frequency standards have been proposed as candidates for the next generation of atomic clocks.

Coherent Optical Transients

One area of laser spectroscopy, coherent optical transients, exploits the temporal coherence of laser light. Gaseous and solid atomic systems can be coherently excited, producing a new and unusual class of nonlinear optical phenomena. Effects such as optical free-induction decay—the coherent emission from atoms excited by a single-laser pulse—and photon echoes—the delayed burst of coherent radiation following excitation by two successive laser pulses—can be applied to study dynamic interactions of atoms in their local environment. Optical free induction provides new ways to study elastic collisions of atoms or molecules that are not in a single eigenstate as in traditional scattering experiments but in a superposition of ground and excited states. Cross sections and other parameters can be determined from measurements of the decay. The close impacts with a perturber that annihilates the superposition state can be visualized classically in terms of separate scattering trajectories, one for each state, resembling state selection in a Stern-Gerlach experiment. Distant impacts or small-angle diffractive scattering where the superposition is largely preserved require a quantum description.

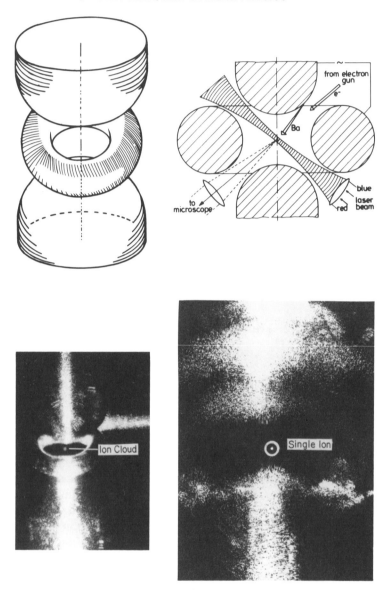

FIGURE 6.2 Trapped Ions. Ions can be trapped in high vacuum using static and oscillating electric fields and viewed by laser light. The experiments can be so sensitive that single ions can be observed under close to ideal conditions of isolation. In this experiment barium ions are formed in the center of the donut-shaped electrode by bombarding barium vapor with electrons. The ions are observed by their fluorescence under laser light. The photograph at bottom left shows the laser light scattered by a small cloud of trapped ions. In the blown-up photograph at bottom right, the light scattered by *one* barium ion can be discerned in the circled region. Laser light can also be used to

Ultranarrow Optical Transitions

Optical free-induction decay of impurity ions in certain solids (the praseodymium ion in lanthanum trifluoride is one example) can display extremely narrow linewidths, 1 kilohertz or less. This is 10^6 times narrower than typical linewidths in solids. These optical transitions are analogs of the Mössbauer effect: the optically excited impurity ion suffers no recoil effect because its momentum is transferred to the lattice as a whole. Furthermore, at cryogenic temperatures there is virtually no second-order Doppler broadening. These systems are prime candidates for studying the interactions that broaden optical transitions, and possibly for establishing secondary optical frequency standards. The method, which is made possible by the use of a dye laser with a 100-Hz linewidth, has been applied to study the optical Bloch equations, the starting point for many theories in quantum optics. It has been found that in intense laser fields the optical Bloch equations must be modified because the radiation inhibits the line-broadening effects of nuclear magnetic interactions. The phenomenon is now understood in terms of a microscopic theory of nuclear magnetic interactions.

Coherent Raman Spectroscopy

The term coherent Raman spectroscopy describes a class of nonlinear optical techniques that are used to study and measure Raman-active modes of molecules. The major techniques are coherent anti-Stokes Raman spectroscopy (CARS) and stimulated Raman spectroscopy (SRS).

In contrast to ordinary Raman light-scattering methods, the signals from CARS and SRS come in the form of strong and highly directional laser beams. As a result, these methods offer tremendous discrimination against undesirable background fluorescence and luminescence.

Coherent Raman spectroscopy works in environments where the

---

cool the ions, reducing the energy-level shifts due to the second-order Doppler effect. Trapped-ion methods are being applied to ultrahigh-resolution optical spectroscopy and to the creation of new types of atomic clocks. The methods are also employed to study collisions and chemical reactions, including reactions at very low temperature, and to study collective motion in charged plasmas. Further discussion is in Chapter 5 in the section on Molecular Dynamics and in Chapter 8 in the section on Precision Measurement Techniques. (Courtesy of the University of Hamburg, Federal Republic of Germany.)

background light level or the need for high resolving power make conventional methods impractical. In combustion diagnosis, the method provides a means of nonintrusively mapping the temperature of a gas and the concentration of its species. CARS studies have been performed in hostile environments such as the combustion chamber of internal combustion engines, in gas-turbine combustors, and even in jet engine exhausts.

Other applications for the coherent Raman techniques include studies of the energy-level distributions that result from optical photodissociation, trace detection of pollutants in gas and liquid phase, the spectroscopy of biological molecules, plasma diagnostics, and applications in the study of hydrodynamic flow.

QUANTUM OPTICS AND COHERENCE

The concept of the photon grew out of Einstein's preoccupation with the statistical nature of light, but it was not until the advent of the laser that the statistics of electromagnetic radiation began to be studied methodically. Light can be observed only in its interactions with matter, however; and so the study of light inevitably encompasses the dynamics of atoms in the radiation field. These subjects collectively form the main body of research in quantum optics and coherence. During the past decade these studies have provided new insights into the statistical nature of radiation and the dynamics of lasers and other quantum systems. In addition, they have opened the way to new methods of measurement and to new quantum devices.

Photon Antibunching

One can observe the arrival of photons at two separate detectors and study the probability that the photons arrive in coincidence. (The light beam is split by a semireflecting mirror, and each half is detected by a separate phototube. The correlations are found from measurements of the coincidence rate.) These are called second-order correlation experiments, since they are sensitive to the product of two intensities. The famous Hanbury-Brown and Twiss experiment determined the diameter of a star from measurements of the second-order correlations in its light. Second-order corrections are usually positive; photons tend to bunch together. It has now been discovered, however, that it is possible to prepare light so that the photons, instead of coming in clumps, are antibunched. More precisely, if a phototube receives a photon, for a short period thereafter it is less likely than otherwise that it will detect a second one.

Antibunching can be observed in light coming from a single atom. (It would not suffice to attenuate a conventional light source, or laser light, for that matter, for that would change only the average arrival rate of photons, not the statistics of the radiation field. Antibunching occurs only when the photons are produced by a nonclassical source.) This has been achieved in practice by using as the light source a single atom that is coherently excited and radiates spontaneously. Other processes, such as harmonic generation or parametric amplification, should also exhibit the antibunching. The novelty of antibunching, however, lies not so much in the realization that one atom can radiate only a single photon at a time but that the statistical properties of the light are different from any that have previously been observed. Antibunching provides an example of light that is fundamentally different from any light previously studied.

Closely related to antibunching is the production of light for which the photon number fluctuations are smaller than random. This has also been observed: it offers the interesting possibility of allowing optical communication with less noise than with a coherent laser beam.

Optical Bistability

An optically bistable system has two stable output states for a given input level of light. Typically, it consists of a nonlinear medium within an optical resonator. Optical bistability was first observed using sodium vapor in a Fabry-Perot étalon, and now there is an expanding class of optically bistable devices. Often the devices are constructed of tiny semiconductor chips whose faces are polished to form a resonator.

Optical bistability provides a new arena for the study of nonlinear systems. Many of the dynamical phenomena that have been studied in lasers, for instance fluctuations and regenerative pulsations, can be observed under far better controlled conditions using optically bistable devices. The transition from ordered to chaotic motion is of particular interest. Such transitions have been studied in hydrodynamic, acoustical, and electrical systems; optical bistability allows the research to be carried out under highly controlled conditions at extremely high speed. The ability to gather data at high speed is of particular value, for it offers the opportunity to study turbulent motion in ways never before possible.

Much of the interest in optical bistability is due to its potential applications to optical computing. A bistable device can serve as a fast memory element; it can be used as an "optical transistor" to amplify small signals at high speeds; and it can be employed as a discriminator, a pulse shaper, an oscillator, or a general logic element. A room-temperature optically bistable device has been created, 5 μm thick and 10 μm in effective diameter, that turns on in a few picoseconds (1 picosecond = 10^{-12} second). These devices can be integrated into larger systems, with numerous applications to optical computing and data processing.

Systematic interest in chaotic phenomena can be traced back to Poincaré's studies early in this century. There has been a great renewal of interest in recent years, and the question of regular versus disordered motion is now a central problem in physics, with important ramifications in mathematics and engineering. Two discoveries, in particular, have contributed to the present interest. One was that attempts to predict long-range weather patterns were inherently limited by the onset of turbulence. The problem of chaos thereby assumes enormous economic significance. The second discovery was the realization of the ubiquitous nature of period doubling in naturally oscillatory phenomena, revealing an important route from regular to chaotic motion.

Optical bistable devices provide a way to study the transition from regular to chaotic motion in reproducible experiments that can be carried out at very high speed. The simplest bistable optical system comprises two mirrors and a nonlinear medium that is operated in a transient mode using pulsed lasers.

The onset to chaos often reveals precursors. In the case of optical bistability these have been discovered to produce short pulses with 100 percent modulation, even in a high-finesse cavity. The technique holds the possibility of new methods for short pulse generation and possibly for optical processing.

Squeezed States

According to quantum mechanics, two canonical variables, such as the position and the momentum of a particle, cannot both be known with great precision; the product given by the uncertainty in one multiplied by the uncertainty in the other must exceed half of Planck's constant. It follows that one variable can be determined accurately only at the expense of large fluctuations in the value of the other. Squeezed states are quantum states that exploit this property. They have become particularly important for the electromagnetic field, in which two oscillatory quadrature (90°-of-phase) components of the field play the role of canonical variables. In a squeezed state one component of the field can be relatively free from fluctuations while the other fluctuates appreciably. This has potentially important applications for optical communications and high-precision measurements, provided that it is possible to encode and decode information in just one quadrature component of the light.

Much effort has been devoted to exploring theoretically the different conditions under which squeezing can be produced. It has been found that squeezing can occur in parametric processes, harmonic generation, phase conjugation, resonance fluorescence, the free-electron laser, and many other circumstances. The practical problem of encoding and decoding information in squeezed light is not without difficulties, but if they can be overcome it would be possible to achieve signal-to-noise ratios in an optical communication channel that go beyond the quantum limit for coherent or laser light.

The problem of detecting gravitational waves with detectors operating close to the quantum limit, where the signal is hidden by quantum noise, has also generated much interest in squeezing. Our ability to make new kinds of astronomical observations may benefit eventually from the use of squeezed states.

Rydberg Atoms and Cavity Quantum Electrodynamics

Any neutral atom in which one electron is in a high-lying energy level is known as a Rydberg atom. These atoms have opened a new area in the study of fundamental radiative processes—cavity quantum electrodynamics.

The interaction between Rydberg atoms and the electromagnetic radiation field scales as n^4, where n is the principal quantum number. For $n = 30$, for instance, the interaction is 10^6 times larger than for ordinary atoms. As a result, the rates at which Rydberg atoms absorb and emit radiation are anomalously large. Thermal radiation at room temperature, usually ignored, is intense for these atoms: it shortens their radiative lifetimes, redistributes the atoms among the various quantum states, and can photoionize the atoms at measurable rates. In addition, thermal radiation can shift the energy levels. The effect is somewhat analogous to the Lamb shift, except that its origin is the real energy of the radiation field, not the virtual energy of the vacuum. The blackbody shift is small but measurable. It needs to be taken into account in the design of the next generation of atomic clocks.

Rydberg atoms are so sensitive to radiation that they provide a natural medium for detecting infrared, submillimeter, and microwave radiation. A number of schemes have been proposed and realized in a laboratory setting. Rydberg atoms can be used to count photons with an efficiency that comes close to the ideal quantum limit. Since absorption is inherently frequency selective, Rydberg atoms can also serve as tuned receivers. In addition, they can be employed in maser amplifiers. Unlike conventional masers, the number of atoms required is small; maser action with only one atom has been achieved.

When an atom is placed in a tuned cavity its radiative behavior is fundamentally altered: the spontaneous radiation rate is enhanced; the lifetime shortened. Rydberg atoms have made it possible to study these effects. If the losses in the cavity are made sufficiently small, a point is reached where the atom no longer decays to the lowest state. The atom and the cavity behave like a pair of coupled oscillators—one atomic, the other man-made. Such a device represents a new entry into the field of macroscopic quantum electrodynamics and provides a unique opportunity to study the transition from reversible to irreversible behavior and the origins of noise.

Cavities not only enhance the spontaneous radiation rate, they can also inhibit it. Simply put, an atom cannot radiate a long wave into a short cavity. From another point of view, the cavity can be viewed as modifying the spectrum of zero-point fluctuations that induce spontaneous emission. If the cavity is mistuned, the fluctuations are removed and spontaneous emission is inhibited. This effect has been seen. It is possible to "turn off" spontaneous emission, leaving an atom in a new type of excited state, a state devoid of radiative damping. The natural linewidth is suppressed, and other radiative interactions such as the Lamb shift are altered.

The ability to observe basic radiative processes with Rydberg atoms offers a new arena for studying electrodynamic phenomena. Although quantum electrodynamics is usually regarded as a highly developed theory, the new experiments suggest that there is a wide body of phenomena yet to be discovered. For instance, it has been found that it is possible to measure the number of thermal photons in a cavity by counting the number of atoms that the cavity can excite. The technique provides, in principle, an absolute thermal radiometer. A scheme has been proposed for cooling a radiation field below the temperatures of its surroundings. Undoubtedly other surprises are in store.

FEMTOSECOND SPECTROSCOPY

A decade ago picosecond optics was in its infancy; today picosecond spectroscopy is a mature field and femtosecond (10^{-15} second) optics is in its infancy. Pulses as short as 9 femtoseconds have been generated; such a pulse contains only 5 cycles of light.

Femtosecond spectroscopy can provide revolutionary insights into the dynamics of molecules and solids, and into chemical reactions, since femtosecond pulses are compared to the characteristic times for all of these. For instance, a molecule typically requires 10^{-14} to 10^{-13} second to vibrate; the time for electrons in a semiconductor to equilibrate after they have been excited can be as short as 10^{-14} second; and proton and electron transfer in molecules can be quicker than 10^{-14} second. The state-to-state dynamical steps in many solid-state and surface processes span an enormous range of frequencies; for the first time femtosecond pulses of light make it possible to observe these phenomena.

Chemical reactions in solutions and in biological systems also take place in the femtosecond regime. Their study is especially important for chemistry since most chemical reactions—organic, inorganic, or biochemical—occur in solutions. The use of femtosecond spectroscopy for the direct, real-time observation of ultrafast relaxations and reactions in condensed-phase chemistry is expected to open new horizons in chemical research. In complex biomolecules the number of energy-transfer paths can be so large that transport and relaxation processes occur on a subpicosecond time scale. Femtosecond spectroscopy can provide a unique means of identifying and studying these primary biophysical events.

Femtosecond optics can also have important applications to fast electronic circuitry and high-speed instrumentation. Femtosecond techniques enable optical pulses to reach a domain inaccessible by electronic techniques. Optical pulses can be used to investigate semiconductor processes that determine the ultimate speed potential of electronic circuitry. For example, high-speed photoconducting pulse generators and sampling gates have been used to measure the electronic input response of gallium arsenide field-effect transistors. The information obtained in such studies will become increasingly important for the design of faster and smaller computers. Ultimately, all-optical modulation and switching techniques, utilizing nonlinear interactions between the ultrashort light pulses themselves, have the potential to go beyond electronics and advance signal-processing speeds into the femtosecond domain.

7

Scientific Interfaces

The boundaries of atomic, molecular, and optical (AMO) physics penetrate far into the neighboring areas of science. Across its borders flows a stream of new techniques and vital data. There is hardly an area of science that has not significantly benefited from these. Geology, geophysics, and planetary physics, for instance, have all been enriched by the maps created with optically pumped magnetometers. Laser surveying, to cite another example, permits monitoring the strains that lead to earthquakes as well as the motion of the continents as they drift and of the moon as it wobbles in its orbit. It also makes possible the high-precision alignment of large particle accelerators and allows tunnels drilled from opposite sides of a mountain to meet exactly.

There is a second form of commerce between AMO physics and the neighboring areas: this is the commerce of basic science itself. In this the boundaries disappear as the unity of science asserts its preeminence. Among the six interface areas that we have chosen to describe here—astrophysics; materials research; surface science; and plasma, atmospheric, and nuclear physics—instances of the underlying unity constantly occur.

ASTROPHYSICS

Most of what we know about the universe comes from information brought to us by photons. To decipher their messages, we must

126

understand how the photons came into existence and the histories of their journeys through intergalactic and interstellar space. From this enterprise we can learn about the early universe and the nature of the astrophysical entities—quasars, galaxies, stars, pulsars, stellar winds, supernova remnants, nebulae, masers, and molecular clouds. The events that produce photons and the processes that modify them during their long journeys lie squarely in the domain of AMO physics. AMO physics is an essential component of astronomy.

AMO physics ranges broadly in its applications to astronomy. The physical and chemical processes that created molecular hydrogen in the early pregalactic universe, which manufactured cyano-octatetrayne in interstellar clouds and propane in the atmosphere of Titan, which bring molecular clouds to the brink of gravitational collapse and trigger star formation, which control the abundance of ozone on the planet Earth, and which determine the radiative losses from stellar interiors, are all part of the body of AMO physics. The unity of AMO physics is manifested in the remarkable diversity of environments in which atomic processes play the crucial role.

Astronomy and AMO physics are mutually dependent. Because the extraordinary physical environments that arise in astronomical phenomena are often impossible to duplicate in laboratories, astronomy provides an extended arena for studying atomic processes. For example, despite the enormous differences in the scales of laboratory and astrophysical plasmas, the fundamental processes are identical. The same mechanism that is believed to be responsible for the lack of electrons at high altitudes in the atmosphere of Jupiter, as revealed by data from the Voyager spacecraft, has been proposed for producing the negative hydrogen ions that are needed to ignite a thermonuclear fusion plasma. The mechanism involves vibrationally excited molecular hydrogen. These molecules are ubiquitous. In the interstellar medium their Doppler-shifted lines signify the shock waves that accompany the birth of stars. Provided that the molecular processes are clearly understood, these lines can provide powerful diagnostic probes of the earliest stages in the evolution of a star.

Interpreting the abundant data of astrophysics demands a deep understanding of atomic, molecular, and optical processes. In addition, it demands a broad data base of atomic and molecular parameters such as transition energies, oscillator strengths, and photon and particle collision cross sections. Providing these data is a major challenge for atomic and molecular physics. Experimental data flow from all branches of the field, particularly from the discipline that has come to be called laboratory astrophysics. These experimental data are vital, but more data are required than the experimental community can possibly provide. Thus, theoretical data are also vital. The need to generate theoretical data for astrophysics motivates a major portion of the theoretical effort in the atomic and molecular community.

The required data base is huge. Several million emission lines are present in the spectra of the Sun and stars. Models of stellar atmospheres have limited success, even in the visible region. Important questions are unresolved: the discrepancy between the predicted and measured solar neutrino fluxes may be due in part to errors in the calculated opacity of the Sun. Stellar explosions, which play a crucial role in the energetics of the galaxy and the formation of new stars, provide another example. These explosions leave a remnant in which elements such as oxygen, sulfur, silicon, iron, and nickel are stripped of all but one or two electrons. The emission lines of these ions, which fall in the x-ray region, can yield not only the element abundances but also the density and temperature of the remnant. Because the relevant atomic data are not available, a comprehensive description of the atomic processes occurring in a supernova remnant has yet to be achieved.

Atomic Processes

Many atomic processes play crucial roles in astrophysics. To cite one example, in recent years the importance of atomic charge transfer in cosmic plasmas has become evident. In interstellar gases composed of elements in their cosmic abundances, ionizing radiation often produces partly ionized plasmas containing some neutral atomic hydrogen. Charge-transfer collisions with the hydrogen drastically modify the ionization structure of the gas. The emission spectra offer a highly specific diagnostic probe of the plasma. Provided the charge-transfer processes are understood, the spectra can serve to establish the coexistence of multiply ionized and neutral material, provide a direct measure of the neutral abundances, and give unique information on the nature of the ionizing source.

Rydberg Atoms

Ionized gases emit photons at all wavelengths. At radio wavelengths, the photons arise from transitions between highly excited Rydberg levels with principal quantum numbers that can exceed 300. Rydberg atoms are large in size and sensitive to disturbances, but the space between stars is nearly empty, and there is room for the Rydberg atoms to survive until they radiate. Because atomic theory can provide a detailed description of the modes for populating and depopulating Rydberg levels, these atoms can be a valuable guide to events occurring in our own galaxy and in external galaxies. They are used to infer the temperature and densities and the hydrogen-to-helium abundance ratio.

Rydberg atoms can now be generated in the laboratory, and their study has developed into a lively subfield of atomic physics. (This is discussed in Chapter 4 in the section on Atomic Structure and in Chapter 6 in the section on Quantum Optics and Coherence.) The prominence that Rydberg atoms assumed in the laboratory in the mid-1970s was stimulated by their discovery in space in the 1960s.

Interstellar Molecules

Transitions between low-lying levels in molecules generate radiation at radio frequencies. Because the photons suffer little attenuation by interstellar dust, the radio emission lines can be seen over large distances. More than 50 interstellar molecules have been discovered. To cite one consequence, the distribution of matter throughout the galaxy has been mapped from the emission lines of carbon monoxide. Radiation from interstellar molecules can extract energy from the interstellar clouds, cooling them to the brink of gravitational collapse. Two of the most fundamental astrophysical processes, nucleosynthesis and the chemical evolution of the galaxy, can be studied by observing the spatial distribution of isotopic molecules such as $^{13}C^{16}O$ and $^{12}C^{18}O$, though the task requires the mastery of the basic molecular chemistry. In order to determine reliable isotope ratios, for example, molecular fractionation must be understood.

Molecular fractionation substantially enhances the abundances of deuterated molecules—molecules in which a hydrogen atom is replaced by a deuterium atom. By joining the theory of ion-molecule chemistry in interstellar clouds with observations of the abundance ratio of the deuterated compounds, the electron density in molecular clouds can be inferred. This density is a critical astrophysical parameter. Gravitational collapse and the fragmentation of molecular clouds to form stars are mediated by free electrons. Despite the deep significance of molecular fractionation, however, one of the essential molecular parameters of fractionation theory remains unknown. As a result, no more than an upper limit can be obtained for the electron density.

Many interstellar molecules are chemically reactive. In the laboratory they exist only as short-lived transient species, difficult to study at high resolution. In some cases—butadinyl and cyanoethynyl are examples—the spectrum can be more accurately measured in space than in the laboratory. In other cases, thanks to recent developments in laboratory techniques and laser technology, laboratory measurements are now superior. Thus, the fine-structure parameters of the reactive neutral atom, carbon-12, were first determined in the laboratory. The results pointed the way to the successful detection of atomic carbon in dense interstellar clouds. A great many other interstellar species with different isotopic constituents await investigation.

Astrophysical Chemistry

Molecular ions occur at crucial points in the ion-molecular schemes that attempt to explain the formation of interstellar molecules. The measured ion abundances provide a sensitive test of the chemical models. Few of the reaction-rate data are available, still fewer at the temperatures prevailing in molecular clouds. The most important reaction pathways may not yet be recognized; the success of the chemical schemes may be no more than an artifact of unreliable data. The very first laboratory experiments on molecular reactions at low temperatures were carried out recently; these may lead to a quantitative description of molecular formation in cold clouds. Molecules have now been detected not only in interstellar clouds but also in the hostile environments of stellar atmospheres and circumstellar shells. It seems likely that they also exist in other astrophysical regimes such as quasars and that they may someday be useful in detecting of x-ray sources and supernovas buried inside dense clouds.

Cosmology

Atomic and molecular processes can provide vital clues to the nature of the cosmos. For example, the distribution of deuterium in the galaxy provides a direct measure of the matter density in the early universe and bears directly on the question of whether the universe is closed or open. The deuterium is detected as a constituent of different molecular species; the chemistry of deuterated molecules must be understood before the total deuterium content can be obtained.

The spectrum of the cyanogen molecule also has direct cosmological significance. From its optical absorption spectrum the relative populations of the two lowest energy levels can be determined, and from this, its temperature. The temperature was found to be 2.8 K. This was the first measurement of the temperature of the universal blackbody background radiation left over from the big bang.

Cosmology can provide unique insights into fundamental atomic principles. From absorption-line measurements toward distant objects at large red shifts, for example, limits can be set on how the fundamental atomic constants can vary in space and time. One result is that the fine-structure constant cannot vary by more than 1 part in 10^{-12} per year.

Space Physics

Astronomy is primarily driven by remote observations, but one component, space physics—the study of the local solar system—is advanced by local experiments with instruments carried aboard spacecraft. The interplanetary medium undergoes violent upheaval where the solar wind collides with the ionized gas in the outer regions of the planets and their satellites, providing a natural laboratory for studying the effects of electric and magnetic fields on the large-scale motions of energetic charged particles. Atomic and molecular physics is essential to understanding the scene. Charged particles are created, scattered, and lost by atomic collisions. Planetary atmospheres respond to solar ionizing and dissociating radiation in a complex array of atomic and molecular processes. The evolutionary paths followed by these atmospheres are affected by escape mechanisms driven by energy transfer in atomic and molecular collisions. The interpretation can be subtle and can lead to unexpected conclusions. For example, the Viking lander on Mars measured a $^{15}N/^{14}N$ isotope ratio 60 percent larger than the terrestrial value, suggesting the operation of a differential escape mechanism for the two isotopes. On Mars, the process of dissociative recombination of ions of molecular nitrogen generates nitrogen atoms with kinetic energies sufficient to escape the gravitational field of the planet. As the isotopes undergo gravitational separation in the atmosphere, the heavier isotope becomes depleted at the high altitudes where escape occurs. A careful accounting of the escape efficiency establishes that Mars once contained a large reservoir of nitrogen gas. Similar mechanisms occur on Venus with a startling corollary. When molecular-oxygen ions recombine on Venus, they produce energetic oxygen atoms that collide with hydrogen atoms and drive the hydrogen out of the atmosphere. The collisions are too weak to drive out the heavier deuterium atoms. As a result, the deuterium/hydrogen ratio on Venus is much larger than anywhere else in the solar system. From the ratio measured by the Pioneer Venus space probe, one can infer that Venus originally had a large abundance of water.

CONDENSED-MATTER PHYSICS AND MATERIALS SCIENCE

Numerous links join AMO physics with condensed-matter physics and materials science. The study described in Chapters 4 and 5 of how x-ray and photoionization spectra of gaseous atoms and molecules evolve as they assemble into liquids and solids reflects one aspect of this interface area. AMO physics has generated experimental techniques ranging from molecular-beam epitaxy and clusters to laser annealing and sputtering. The impact of AMO physics on surface science—one of the liveliest areas in solid-state physics—is so large that it is described separately in the next section. In this section, we describe three activities: light-scattering spectroscopy, metal clusters, and the creation of spin-polarized quantum fluids.

Light-Scattering Spectroscopy

The extraordinary spectral purity of gas lasers has made them an important source of radiation for the observation of thermally excited fluctuations and of fluid flow in condensed-matter systems. The interaction of the laser radiation with spontaneous molecular motion produces spectral broadening or frequency shifts in the scattered light that generally range from 1 to 10^5 Hz. (The frequency of visible light is about 10^{15} Hz.) The accurate resolution of such small-frequency shifts has become possible using the techniques of optical mixing spectroscopy. These techniques represent the successful extension of heterodyne and homodyne detection methods, long employed in radio-frequency and microwave spectroscopy, upward into the optical frequency domain. As a result of these advances, a new form of spectroscopy, known variously as quasi-elastic light-scattering spectroscopy, photon-correlation spectroscopy, or intensity-fluctuation spectroscopy, has emerged and been applied to a wide range of fundamental and applied problems in physics, chemistry, biology, engineering, and medicine.

Order-Disorder Transitions: In physics, light-scattering spectroscopy has provided many of the basic determinations of the critical exponents for the divergences of the equilibrium and transport coefficients of pure fluids and binary mixtures near their order-disorder phase transitions. The profound nature of the theoretical ideas needed for the exploration of these and related experiments in magnetic systems culminated in the creation of the renormalization group theory, one of the major achievements of modern condensed-matter theory. Quasi-elastic light scattering has been the principal experimental tool used to investigate the hydrodynamic modes and phase transitions of liquid-crystal systems, a field of high current interest. Light-scattering spectroscopy has been used to study the relaxation of density fluctuations in gases, the propagation of elementary excitations in liquid helium, and the development of soft modes in the phonon spectrum in solids near phase transitions. It is a principal experimental means of investigating the transition to turbulence (or chaos) in hydrodynamics and has been used to determine the value of important universal numbers in the theory of strange attractors.

Applications to Chemistry, Biology, Engineering, and Medicine: In chemistry, the method is widely used to obtain important microscopic information on the fundamental interactions between amphiphillic molecules, which self-assemble to produce well-defined geometrical structures: micelles, microemulsions, vesicles, and bilayers. These structures are fundamental constituents of the living cell and are of great importance in a wide variety of industrial chemical processes. Quasi-elastic light-scattering spectroscopy has been used to discover scaling phenomena in polymer solutions and to examine the moments of polymer cluster size distributions near the sol-gel transition. In polymer gels it has been used to discover a rich variety of hitherto unexpected first-order phase transitions. The latter phenomena are potentially promising for the development of mechano-chemical, mechano-electrical, and electro-optical devices.

In biology, quasi-elastic light-scattering spectroscopy has been used to determine quickly and accurately the diffusion coefficients and hence the size and degree of self-association of a wide variety of biological macromolecules including proteins, viruses, and antibody-antigen complexes. These studies have been used to characterize accurately the precise form of the Coulomb and van der Waals interactions between polyelectrolytes in ionic solutions. The method is also used in studies of colloid stability, ordering, and flocculation.

Light-scattering spectroscopy has led to the opening of the broad field of laser Doppler velocimetry, which permits noninvasive measurements of fluid flow in situations ranging from aircraft wake velocity fields to the in vivo determination of blood velocity in the human retinal vasculature.

Atoms in Solids: The theoretical insights and techniques developed to describe atomic and molecular structure are being applied to problems of condensed-matter structure. Two examples are the description of electronic states of impurity ions and atoms in crystals, for instance Ca^+ in crystalline LiCl or H in amorphous silicon, and the calculation of the band-gap energies of insulators and semiconductors.

The approach is straightforward, at least in principle. In the independent-electron approximation, the exchange-correlation interaction is represented by a one-electron local potential. The variational wave function is represented as a linear combination of atomic orbitals, just as in molecular-structure calculations. Carrying out such a calculation is a formidable task. But by representing the exchange-correlation interaction with a simple local density-dependent exchange potential (ignoring correlation entirely), and using a series of Gaussian functions to describe the atomic orbitals, the theory becomes tractable, even for disordered systems for which the standard band-structure methods are not applicable. Density functional theory has established that the one-electron density uniquely defines the ground-state energy of a system, but it is surprising how well the local-exchange theory works.

A serious difficulty is that the theory fails to predict accurately the band gap, the energy separation between the uppermost valence levels (top of the valence band) and the lowest conduction levels (bottom of the conduction band) of insulators and semiconductors. Recently, dramatic improvement in the theoretical predictions was obtained by making a simple self-interaction correction to the total energy, thus bringing the local-exchange theory more in line with true Hartree-Fock theory, in which the correction is implicit. It has also been shown that the correction leads to much improved energies for isolated atoms. While correlation effects may still prove to be important in some circumstances, the self-interaction correction appears to be a significant improvement in the quality of the theory.

Clusters

Chemical and physical processes often occur in a state of aggregation that lies midway between a dilute gas and condensed matter. The entities of this state are aggregates of small numbers of atoms or molecules called clusters. The properties of clusters are intermediate between those of single atoms or molecules and those of solids or liquids. Many of the processes that occur in the cluster regime are important to technology and industry and to environmental issues. These include catalytic reactions; the formation of fog, smog, and aerosols; and the formation of particulates in combustion reactions. Clusters play a role in solution chemistry because they can retain their identity even in the liquid phase.

In contrast to the detailed spectral information that exists for atomic and molecular dimers, information on the electronic properties of trimers and heavier clusters is scarce. Recently, the electronic absorption spectrum of the sodium trimer was determined over the complete visible region of the spectrum in a two-photon photoionization experiment. The experiment provided the first unambiguous measurement of the absorption spectrum of a gas-phase triatomic metal cluster. At present, spectral or structural information about gaseous clusters beyond the trimer are lacking. These data are critically needed to provide the link between the dimer and the bulk phase. The one continuous property known today for heavier clusters, from 2 to 15 atoms, is the photoionization potential. The earliest measurements of photoionization potentials were on alkali clusters; however, more recently photoionization thresholds as a function of cluster size have been reported for other species including rare-gas clusters, metal clusters, and a few molecular clusters such as $(CO_2)_n$, $(CS_2)_n$, and $(H_2S)_n$.

Laser-induced fluorescence has been used to determine the spectra of dimers of large organic molecules. These studies provide information on the energetics of cluster formation, for instance the bond dissociation energy, and information on the transfer of energy between the two moieties of the dimer via the weak van der Waals bond. Other studies have determined the spectra of an organic molecule bound to an increasingly large number of rare-gas atoms. Since the rare-gas atom acts as a weak perturber of the energy levels of the host molecule, these studies approximate matrix isolation studies, allowing the detailed determination of the effects of the matrix on the spectra of the host molecule. One can also approach the cluster region from the solid state. Here the goal is the size at which the collective properties of the solid disappear as the particle diameter is reduced. Experimental data have been reported for melting point, superconductivity, valence-band narrowing, photoelectric yield, plasmons, Mie optical absorption, magnetic moments, Compton profile, superparamagnetism, far-infrared absorption, specific heat, and crystallographic structure.

Clusters can also be used to study surface physics, as described in the following section on Surface Science.

Ultranarrow Optical Transitions

Within the last few years it has been found that certain optical transitions of impurity ions in solids (the praseodymium ion in lanthanum trifluoride is one example) display extremely narrow linewidths, 1 kilohertz or less. These optical transitions, the zero-phonon lines, are optical analogs of the Mössbauer effect: the optically excited impurity ion suffers no recoil effect because its momentum is transferred to the lattice as a whole. Furthermore, at cryogenic temperatures there is virtually no second-order Doppler broadening. These systems are prime candidates for studying the interactions that broaden optical transitions and possibly for establishing secondary optical-frequency standards. The method has been applied to study the optical Bloch equations, the starting point for many theories in quantum optics. It was found that intense laser fields can inhibit the line-broadening effects of nuclear magnetic interactions. The phenomenon has spurred reconsideration of microscopic theories of nuclear magnetic interactions.

This research has provided the first experimental test of the optical Bloch equations, the equations of motion that were initially devised by F. Bloch to describe nuclear magnetic resonance. These equations are widely applied in quantum optics and laser spectroscopy, particularly in gases and liquids; they are the starting point for work in these fields. However, in solids it has been discovered that they fail because the laser field amplitude increases because of a coherent averaging effect that reduces the optical linewidth. A microscopic quantum theory, a modified form of the Bloch equations, has been devised to deal with this situation.

Spin-Polarized Quantum Fluids

All forms of matter solidify at sufficiently low temperature except for one class of systems—the quantum fluids—which remain in liquid or gaseous states as the temperature approaches zero. Within the past few years two new quantum fluids have been created using techniques from AMO physics: spin-polarized gas ^{3}He and spin-polarized atomic hydrogen.

Because ^{3}He has a total spin of one half, the atoms obey the Pauli principle and there is an effective repulsion between them when their nuclear spins are parallel. As a result, diffusion, viscosity, and thermal conductivity of the gas all depend on the nuclear polarization. The density of the liquid is expected to depend on the polarization, decreasing slightly as the polarization is increased because of the Pauli repulsion. The bulk properties such as magnetic susceptibility and thermal conductivity should also be correspondingly altered.

A gas of polarized ^{3}He at cryogenic temperatures has been produced using atomic optical pumping techniques. A color-center laser provides the intense light needed to pump the atoms. The ^{3}He gas is remarkably stable—the nuclear polarization can last from half an hour to days. The properties of the gas are just now starting to be investigated. Potential applications include polarized ^{3}He sources and targets for nuclear physics, sources for polarized electrons and molecular ions, and use as a neutron spin filter.

The second new quantum fluid is spin-polarized hydrogen. Under normal conditions hydrogen atoms join to form molecules in a violent reaction, but, if the electron spins are all kept parallel, molecules cannot form. The system is predicted to remain gaseous at arbitrarily low temperatures.

Spin-polarized hydrogen is formed by cooling the atoms to below 1 K and polarizing their spins in a high magnetic field. Superconducting magnets and dilution refrigerators are key elements of the method, but the basic techniques came from AMO physics: magnetic deflection using what might be called a "super-Stern-Gerlach" technique and the use of liquid helium wall coatings to prevent surface recombination and relaxation of hydrogen.

The transport properties of spin-polarized hydrogen are expected to depend on the nuclear polarization. It has been found that molecular recombination of the gas can be initiated and controlled by changing the direction of the nuclear spin, providing the first example of a chemical reaction that can be controlled by changing the orientation of a nucleus.

In addition to its interest as a quantum fluid, spin-polarized hydrogen promises to have useful applications in many fields. To mention a few: The new techniques may lead to a type of hydrogen maser that is superior to existing atomic clocks. The techniques of spin-polarized hydrogen are being adapted to the production of polarized proton sources and targets for particle and nuclear physics. Other applications range from the creation of slow atomic hydrogen beams for super-precise spectroscopy and hydrogen scattering experiments to the production of polarized deuterons for the proposed use of polarized nuclei to obtain energy-producing fusion plasmas under less extreme conditions than previously contemplated.

SURFACE SCIENCE

Surface science deals with questions ranging from the charge distribution and vibrational structure in the surfaces of metals, insulators, and semiconductors to the dynamics of adsorption and the chemistry of interfaces. The field is advancing rapidly under the combined impact of new experimental techniques and growing interest in the solid-surface region. The research bears on the structure of surfaces, on two-dimensional systems, and on physical and chemical surface processes. It is central to the whole subject of catalysis. Other applications include electronic materials and processes, thin-film physics, corrosion, aircraft drag, and lubrication.

Techniques from AMO physics such as laser spectroscopy and molecular-beam scattering are helping to revolutionize surface science. (See Figure 7.1.) Often, surface science is carried out in AMO laboratories. Here are some of the contributions.

Molecular-Beam Surface Scattering

Neutron scattering revolutionized the study of solids by providing a probe with a wavelength well matched to the crystal periodicity. Molecular beams are now providing a comparable probe for surfaces. The new probe is the helium atom, for the wavelength of helium atoms in a supersonic beam is commensurate with the spacing of particles on surfaces. Furthermore, helium atoms, in contrast to neutrons, do not penetrate the bulk material; they interact only with the surface. Molecular-beam surface scattering employs two techniques developed in the course of studying molecular collisions: intense monoenergetic beams of helium and detectors of extraordinary sensitivity, better than one billionth of a billionth of atmospheric pressure. Surface scattering can reveal atom-surface interactions and forces that govern energy accommodations and adsorption. By analyzing the diffraction data with powerful theoretical inversion techniques, surface charge densities can be determined.

One of the most dramatic developments in surface scattering is surface phonon spectroscopy, which employs angle- and velocity-resolved inelastic helium scattering to study surface vibrations. The technique provides a high-resolution probe of surface vibrations that is complementary to electron energy-loss spectroscopy. Advantages include excellent surface sensitivity and energy resolution in the submillielectron-volt range. Because the momentum of the helium is well matched to that of surface phonons, the entire surface Brillouin zone can be studied, including the short-wavelength phonons that are sensitive probes of surface force constants and surface structure. Surface vibrations have been observed on metals, metal oxides, alkali halides, and semiconductors. The method should be useful for adsorbate-covered surfaces and may even reveal lateral interactions within adsorbate layers. Such interactions are important for understanding physisorption and chemisorption, including two-dimensional phase transitions. Recently, a series of experiments involving epitaxially grown thin films of xenon supported on silver have begun to reveal how the lattice dynamics of thin films evolve into those of a bulk crystal on a layer-by-layer basis.

Metal Clusters

Using lasers it is possible to vaporize a metal target within a supersonic nozzle, creating an intense ultracold beam of small clusters of the bare metal. The technique generates clusters in sizes from 2 to 200 metal atoms. It works just as well for tungsten, the highest boiling material known, as it does for aluminum. (See Figure 7.2.) By using targets composed of alloys or sintered mixtures of metals, unusual clusters can be prepared. Furthermore, the clusters can be reacted with molecules such as carbon monoxide, hydrogen, or nitrogen to prepare cold beams of the chemisorption product.

Metal cluster research is likely to have a major impact on surface science, particularly on the study of heterogeneous catalysis. Bonding within the metal cluster is expected to be so cohesive that essentially all the interactions of a cluster with its surroundings are determined by the surface properties. Even for clusters containing as many as 100 atoms, over 50 percent of the atoms lie on the surface. A cluster's properties are expected to be radically different from the bulk metal and from the isolated atom; this difference is crucial to the operation of many important industrial catalysts.

Little is yet known about metal clusters. We do not know the nature of the metal-metal bond or how it changes as one adds additional metal atoms. Many questions need to be answered. How big must a cluster be to behave like a metal with truly itinerant electrons? How does the band structure of a metal develop as atoms are added to the cluster? Most importantly, what is the chemistry of the metal cluster surface? Some important chemical reactions occur only on metal surfaces, for instance, the Fischer-Tropsch conversion of hydrogen and carbon monoxide to form hydrocarbons. Hardly anything is yet known about

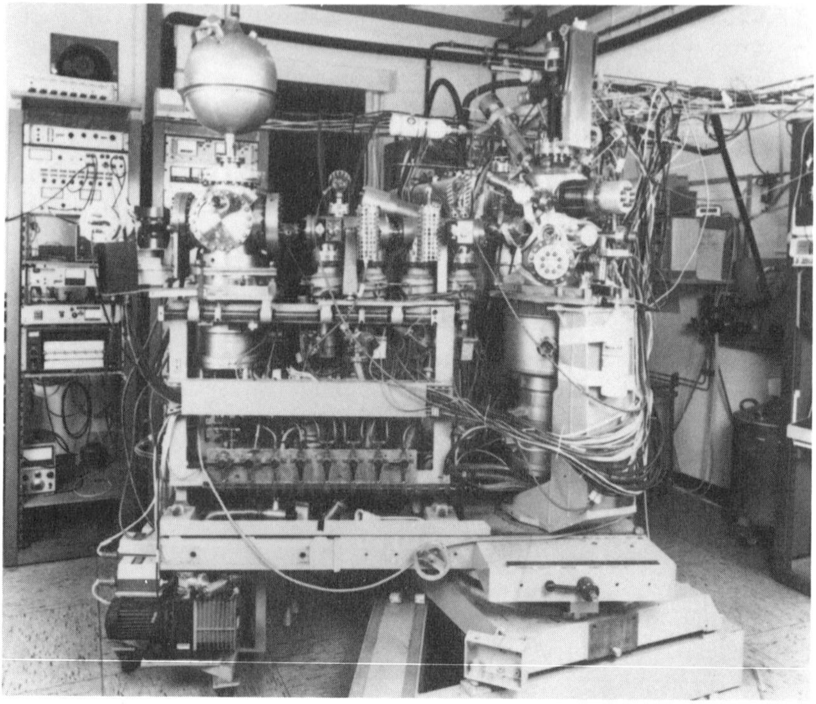

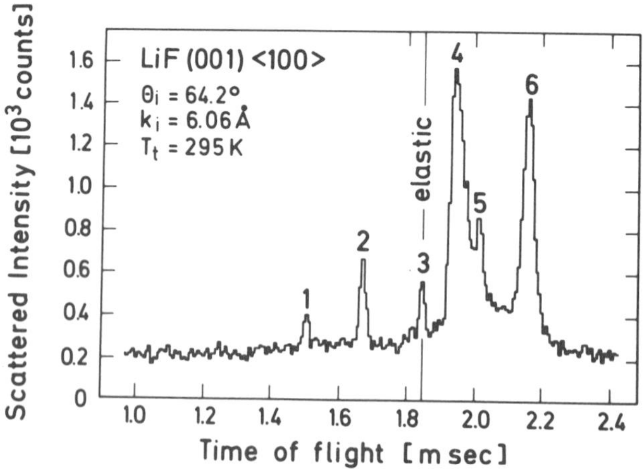

FIGURE 7.1 Surface Scattering with a Supersonic Helium Beam. An atomic beam of helium constitutes a powerful probe for studying surfaces. In fact, it is believed that the role of helium scattering in the study of surfaces may be comparable with the role played by neutron scattering in the structure of solids. An intense monoenergetic flux of atoms is required; this is provided using supersonic-beams methods developed from research

the details of such reactions; clusters may provide the key to understanding what happens.

Studying Surfaces with Laser Light

Laser light makes it possible not only to study surfaces in ways never before possible but also to change the surface in new ways. Short intense laser pulses can reveal dynamical surface phenomena; coherent UV light can produce new types of nonlinear surface effects when it strikes adsorbed molecules. Laser light can trigger chemical changes on surfaces, in the substrate, and in the overlying gas.

How a surface affects a photochemical reaction depends critically on whether the molecule decomposes immediately in the light or whether the reaction takes place slowly; with pulsed laser techniques the two alternatives can be distinguished. Chemical reactions of particles adsorbed on a semiconductor can be triggered in a controlled fashion—essentially catalyzed—by using laser light to generate electron-hole pairs within the material. The holes drift to the surface and trigger the reaction. If the process can be made to occur at a gas-solid interface it could provide an immensely useful new catalytic technique. The surface-sticking coefficient for vapor-phase metal atoms can change by decomposing a thin adsorbed layer of metal alkyls with laser light, offering for the first time a precise way of controlling the interchange of energy between a gas and a surface. Finally, photoreactions on the surface can trigger the growth of new materials with novel properties. The technique has important applications to semiconductor electronics and to electro-optics.

A dramatic discovery from the study of surfaces with laser light is that Raman spectra on surfaces can be enhanced by a magnification of the local optical electric field. The enhancement can be enormous—as much as a factor of 10^6. A typical experiment uses green laser light to illuminate a silver surface containing microscopic spheres or ellipsoids. These particles exhibit a plasma resonance that magnifies the electric fields in their vicinity. The plasmon resonances are of considerable interest in their own right. The technique provides an extremely

---

on molecular scattering. The data show clearly resolved structure in the speeds of atoms scattered from a lithium fluoride crystal. The spacing of the peaks provides detailed information from which the structure and motions of atoms on the surface can be determined. (Courtesy of Max-Planck-Institute for Fluid Dynamics, Göttingen, Federal Republic of Germany.)

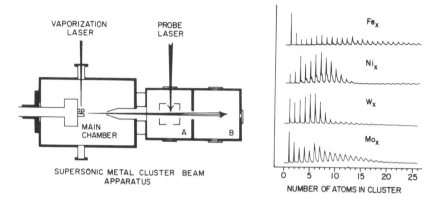

FIGURE 7.2   Metal Clusters. Clusters are small groups of atoms or molecules in a state of matter intermediate between a dilute gas and condensed matter. Supersonic beams of metal clusters are made by vaporizing the metal in an intense pulsed supersonic beam of helium. Using a high-power pulsed laser it is possible to vaporize even the most refractory metals. The data show mass spectra for clusters of iron, nickel, tungsten, and molybdenum. Most of the atoms in the clusters lie on the surface, even for clusters as large as 100 atoms. Because the physical and chemical properties of the clusters are dominated by surface phenomena, supersonic metal atom clusters provide an important new arena for surface science. The technique is particularly valuable for the study of catalysis, a subject of high scientific interest with potentially important applications in chemistry, in manufacturing, and in energy programs. (Courtesy of Rice University.)

sensitive tool for studying surfaces because small quantities of adsorbates in the vicinity of the metal particles are easily detected.

It has been found that silicon can reproducibly change from a crystalline solid to the amorphous state and back again to a crystal with successive picosecond laser pulses. Picosecond-pulse probes have been used to study the solid-state plasma that is formed in silicon when it is illuminated by a pulse of light from a short-wavelength laser. Such studies provide a new and novel way to study the dynamics of crystal growth and to understand the mechanisms that underlie the many applications of laser annealing.

These are but a few examples of how laser light can be used to study surfaces and surface chemistry. The opportunities are great, and the field is growing rapidly.

The Role of Atomic, Molecular, and Optical Data in Surface Science

In addition to contributing experimental and theoretical techniques to surface science, AMO physics provides basic data that are essential to the interpretation of much surface research. For instance, Auger electron spectroscopy and x-ray photoelectron spectroscopy, widely used techniques for determining surface chemical composition, demand extensive data from AMO physics: ionization cross sections for electron and photon excitation, electron binding energies, and Auger transition probabilities. Other data are needed to relate the Auger lineshapes to the chemical states of atoms in molecules and to interpret the probabilities of multielectron excitations in core-level spectroscopies.

Laser studies of surfaces require multiphoton ionization probabilities for atoms and molecules, fluorescence lifetimes and probabilities, and photoabsorption cross sections. Vibrational and rotational emission and absorption spectra for hot molecules are also needed.

Stimulated desorption studies and sputtering spectroscopy require impact cross sections for ionization, branching ratios for dissociative ionization processes in small molecules, and spectroscopic data on ions and highly excited neutral species.

PLASMA PHYSICS

Plasmas are systems of ionized gas whose behavior is determined in large measure by collisions of electrons and ions with each other and with any neutral material that may be present. Plasmas range in physical conditions from hot fully ionized magnetohydrodynamic plasmas occurring in planetary and interstellar environments to various experimental devices. As our understanding of the physics of plasmas has deepened, the importance of atomic processes in plasmas has become apparent. Atomic processes play a basic role in the creation of most plasmas. Neutral-particle injection into magnetically confined plasmas is used to raise the plasma temperature and bring it to ignition. Atomic processes are crucial diagnostic probes of the physical conditions in a plasma. The temperature, density, and fraction of different ionization stages can be derived from measurements based on atomic physics. Atomic processes ultimately destroy the plasma after the initiating source is terminated. The ionization is lost by atomic recombination, and the gas cools by atomic radiative processes.

There is an enormous variety of plasmas in nature: most of the universe is one form of plasma or another. A multitude of atomic and molecular processes occur in plasmas, and there is an expanding need for reliable data on these processes. National plasma facilities should be available to AMO physicists so that the relevant atomic experiments can be carried out. Results of the experiments will be an important element in the physics of dense plasmas. For example, the effects of high electron densities on atomic and molecular processes are surely important but largely unknown.

Laboratories with tokamaks, mirror machines, and other devices intended for the controlled thermonuclear fusion program have become important centers of atomic-physics research. The plasmas produced in these machines contain an abundance of radiation, electrons, and ions in many states of ionization and with energies that are far from uniform. All of this complexity makes a plasma not only a fertile ground for applications of diagnostic atomic-physics techniques but also a valuable source of new information about the interactions of radiation and particles. In such a complicated environment, it is, of course, not easy to study specific individual interaction processes, but observations in plasmas have nevertheless yielded rewarding results for such areas as the rich spectroscopy of satellite lines of highly ionized atomic species. Conversely, electron-ion beam collision experiments have recently substantiated the main features of the dielectronic recombination process that commonly occurs in plasmas. In dielectronic recombination a free electron is captured by an ion, with the simultaneous excitation of an atomic electron. Thus it is evident that atomic physics and plasma physics support each other significantly at their common frontier.

Weakly ionized plasmas containing molecular gases raise a new set of questions concerning the influence of atomic and molecular processes on the evolution of the plasma. Molecular plasmas, energized by some external source, can by virtue of internal excitations modify the course and change the products of molecular reactions.

Fusion research provides a major arena for the interplay of atomic and plasma physics, as described in Chapter 8 in the section on Fusion.

ATMOSPHERIC PHYSICS

Electrical and chemical processes in the atmosphere effect us vitally: they govern massive climatic patterns through their influence on the energy flow from Sun to Earth and from Earth to space; they determine the ultimate fate of industrial pollutants; they control the quality of local and worldwide radio communication. Atmospheric physics, which attempts to understand the complicated web of physical and chemical processes in the atmosphere, draws heavily on AMO physics for vital data and for theoretical and experimental guidance.

At its most elementary level, atmospheric physics deals with the physical processes that occur when our atmosphere is subjected to radiation, electromagnetic and corpuscular, from the Sun. A complex sequence of atomic and molecular processes determines the distribution of the absorbed solar energy into ionization, dissociation, luminosity, and heating; these processes drive dynamical and plasma interactions and are modified by them. The importance of each individual process depends on the location in the atmosphere and on the time: latitudinal, diurnal, seasonal, and solar-cycle variations are all substantial.

The quality with which radio waves propagate is determined by a balance between photoionization and recombination in a large electrified region of the atmosphere. Sunlight creates atomic ions; these recombine after being converted to molecular ions by ion-molecule reactions. A successful model has been constructed for the chemistry governing recombination, though important questions remain about the role of metastable species and vibrationally excited neutral and ionic molecules. A separate group of physical processes governs the history of the photoelectrons in the atmosphere. Initially energetic, these lose their energy first by exciting and ionizing the atmospheric constituents and finally through elastic collisions with the ambient electron gas. The ambient gas is preferentially heated, and its temperature rises above that of the neutral atmosphere. The hot electron gas is cooled by excitation of the fine-structure levels of atomic oxygen and by excitation of the rotational and vibrational levels of molecular nitrogen and oxygen.

The processes that lead to the day and night airglow of the atmosphere have been broadly categorized, but they are not understood in detail. This is also true for the atomic and molecular processes that follow auroral bombardment at high altitudes. Light from the aurora is a potentially powerful diagnostic probe of the exciting source and of the acceleration mechanism that appears to occur. High-latitude auroral events and polar-cap absorption events produce thermal gradients in the high atmosphere, driving the upper-atmosphere winds. They modify the composition of the atmosphere, and they may be related to climatic variations.

The chemistry of the mesosphere and stratosphere has undergone rapid development, particularly since it was recognized that the release of fluorocarbons into the atmosphere could attack the ozone layer. Further studies of the molecular processes are needed to understand the potential hazards from fluorocarbons and other pollutants. The penetration of solar radiation is not yet adequately known, nor is the intensity of ultraviolet radiation at low altitudes.

The terrestrial atmosphere has evolved markedly since the formation of the planets owing to many influences, including life. Molecular physics is crucially involved in the attempts to reconstruct the history of the early atmosphere as it responded to changes in solar luminosity and to understand the interactions today of the physical and biological processes that are determining the future of the atmosphere. The effects of an increase in the abundance of carbon dioxide are of crucial importance to our future.

Basic problems presented by atmospheric science often have immediate consequences. For example, the Space Shuttle was found to glow in the dark even at altitudes as high as 300 km (200 miles). The origin of the glow has not been discovered: it may be produced by collisions of the oxygen atoms of the atmosphere with material on the surface of the spacecraft. The emitting species appears to be molecular in character, but its identity is uncertain. A similar glow observed on orbiting satellites has been attributed to the hydroxyl radical, but the limited data on the Space Shuttle glow suggest that a different molecule must be responsible. It is essential to identify the source of the glow, not least because of its potential impact on the durability and effectiveness of instruments in space such as the Space Telescope.

NUCLEAR PHYSICS

Atomic and nuclear physics are closely related. Here we focus on three areas of contact. The first is the role of atomic spectroscopy in measuring the fundamental static characteristics of nuclear states. This has been an indispensable tool of nuclear physics for decades but now contributes more information than ever. The second is the use of atomic techniques to provide polarized nuclei for sources and targets in nuclear experiments. The third is the study of the dynamical interactions between nuclei and their atomic environments. The physics at this interface between the two disciplines, still in its infancy, has already provided nuclear physics with new insights and tantalizing clues.

Optical Studies of the Nucleus

The mass, size, shape, and internal structure of the nucleus at the center of each atom slightly alter the positions of the atomic energy levels. These energy-level shifts can be found from careful determinations of the optical spectral lines of the atoms. The name "hyperfine structure," given to a large class of these effects, emphasizes their intrinsic smallness, but atomic spectroscopy stands out in physics through its extreme precision, and these tiny effects are accessible to measurement and analysis. For example, although nuclear diameters increase with the mass of the nucleus, even the largest nuclei are about one hundred thousand times smaller than most electron orbits in the atom. Nevertheless, atomic spectroscopy provides a powerful tool for accurate measurement of very small changes in the nuclear radius.

The complex nature of the nucleus is revealed to the atomic electrons that surround it through the electric and magnetic fields that arise from the nuclear protons, neutrons, and pions and, at a more fundamental level, the quarks. These electromagnetic properties of the nucleus can be parameterized in terms of electric and magnetic moments that describe the size, shape, and charge and current distributions of the nuclear constituents. The spins, the magnetic dipole moments, and the electric quadrupole moments of a wide variety of nuclei can be found from measurements of the hyperfine interaction made by two atomic techniques: high-resolution optical spectroscopy and atomic-beam magnetic resonance. Such data provide valuable input to nuclear theorists, who use them to test the still-evolving theories of nuclear structure.

Hyperfine measurements have been fundamental to our understanding of the most basic properties of the ground states of stable nuclei. It has now become possible to apply these techniques to a large class of unstable and short-lived nuclei, and the horizons of the field have expanded dramatically. In addition to the ground states of vast numbers of nuclear species, there are many relatively long-lived excited states whose size, shape, and moments can now be measured. Every bit of such information adds a further constraint on possible theories of nuclear structure.

To appreciate the potential of this new development, one only has to note that while there are about 190 stable or naturally occurring isotopes, there are close to 1600 known unstable nuclei, and it is expected that many more will yet be discovered. Lasers, with their sharp wavelengths and high intensities, have been the principal tool opening up new vistas in the study of nuclei by atomic spectroscopy. The field has been enormously enhanced by the installation of lasers on-line at nuclear reactor and accelerator facilities, where the unstable nuclear isotopes are produced. (See Figure 7.3.)

Systematic studies of chains of isotopes, especially at the Isotope Separator On-Line (ISOLDE) facility at the European CERN laboratories, are providing vital information for nuclear-structure physicists. Techniques from atomic physics are extensively employed to study the sizes and shapes of the nuclei, their spins, and their electric and magnetic moments. These techniques include atomic-beam magnetic

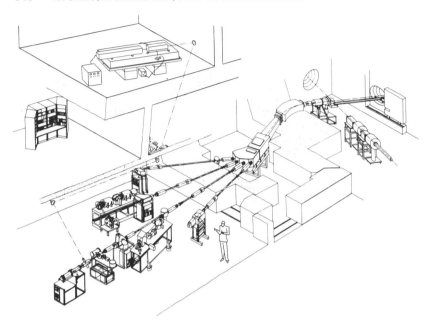

FIGURE 7.3.   Atomic Physics at ISOLDE. The earliest evidence that atomic nuclei possess spin and magnetic moments came from atomic physics, and as new experimental techniques have been developed, the range and precision of these nuclear studies have increased steadily. The drawing illustrates a number of atomic experiments on nuclear properties being carried out at ISOLDE, the on-line isotope separator at CERN (Geneva). Devoted to research in nuclear physics, ISOLDE is capable of producing essentially any isotope. The isotopes are formed by bombarding a target with a 600-MeV proton beam from a proton synchrocyclotron (right-hand side of drawing). The radioactive ions are accelerated and mass selected by a bending magnet and then distributed to the experiments by a "switchyard." Among the equipment shown on the experimental floor are an atomic-beam magnetic resonance apparatus and a setup for laser spectroscopy (the laser can be seen two floors above). Optical pumping is also employed. A similar experimental facility is being developed at the University Isotope Separator (UNISOR) at Oak Ridge. (Courtesy of CERN, Geneva, Switzerland, and Laboratoire Aime Cotton, Orsay, France.)

resonance, optical pumping, and laser spectroscopy. Nuclei with lifetimes as short as 10 milliseconds have been studied. Out of this research have emerged discoveries such as shape staggering, in which adjacent isotopes alternate between oblate and prolate forms, and shape isomerism, in which a nucleus with two nearby excited levels can assume widely varying shapes. These discoveries have played an important role in the development of nuclear models.

As laser spectroscopy continues its rapid advances one can expect corresponding advances in our ability to obtain spectroscopic information about nuclear states. It will become possible to measure the properties of extremely rare and short-lived nuclear states, including highly excited collective states with very high spin values. The remarkable recent progress made by atomic physicists in trapping ions and atoms for long periods of time is certain to be exploited for further high-precision measurements of nuclear moments.

Polarized Nuclear Sources

Nuclear physics relies on techniques from atomic physics for producing the spin-polarized projectiles and target atoms that are being used increasingly in nuclear-reaction experiments. Nuclear physicists require the most intense available beams of nuclei with their spins oriented in a particular direction in space, rather than being randomly oriented. A number of different atomic methods are used for producing polarized nuclei, such as protons, deuterons, ^{3}He nuclei, and lithium nuclei. The spins of nuclei are aligned through their magnetic moments, which offer a "handle" for using a magnetic field to rotate and orient them. However, the nuclear moment is only about one thousandth as large as the magnetic moment of the electron. Therefore, almost all nuclear polarization schemes rely on first polarizing one of the atomic electrons that surround the nucleus. The hyperfine interaction, which couples the magnetism of the electron to the magnetic moment of the nucleus, can then be used to align the nuclei. (Two recent advances in the production of polarized protons and ^{3}He are described earlier in this chapter in the section on Condensed-Matter Physics and Materials Science.) In this important area of ion-source technology it is essential to know the atomic collision cross sections and other atomic parameters that determine the efficacy of proposed new polarization mechanisms.

Dynamics at the Atom-Nuclear Frontier

Atomic and nuclear physics have a common frontier that has recently become the site for research into questions that previously could never be addressed but that now, thanks to experimental and theoretical advances, we can hope to answer. What happens to the electrons in an atom when a nuclear particle, such as a proton or a heavy ion, penetrates through the atomic electron shells on its way into or out of the nucleus, where it initiates a nuclear reaction? How does the course of the nuclear reaction—and in particular its duration—

affect the atomic electrons? Can lessons for nuclear physics be learned by studying these atomic effects? Or, conversely, can useful atomic information be gained? In a number of accelerator laboratories that use all types of accelerators from tandem Van de Graaffs to the most advanced heavy-ion linear colliders, the effects of nuclear reactions at millielectron volt to gigaelectron volt energies on the participating atoms have come under study. The research has begun to yield valuable information on nuclear reactions and atomic-collision processes.

One of the most direct measurements at the atomic-nuclear frontier is the determination of the lifetime of a compound nucleus, before it comes apart again with re-emission of a proton. On its way into or out of the nucleus, the projectile may knock an inner-shell electron out of its orbit, leaving a vacancy, which eventually leads to the emission of an x-ray photon. From such x-ray measurements, and with the aid of results from calculations in atomic-collision theory, it is possible, from purely atomic observations, to determine the lifetime of the compound nucleus, using the atom as a clock. Lifetimes in the range from 10^{-16} to 10^{-18} second have been measured by such atomic techniques. Lifetimes of even much shorter-lived nuclear states (10^{-18} and even 10^{-20} second) can now be determined by crystal blocking techniques that exemplify the overlap of nuclear physics with both condensed-matter and atomic physics.

Finally, we note here a new development: an experiment that is being carried forward at a laboratory of particle physics but that represents a confluence of particle, nuclear, and atomic physics. This is the attempt to make protonium, an atom composed of a proton and an antiproton. The antiproton storage ring at CERN produces enough of these particles to provide the hope of making protonium, using hydrogen negative ions as the source of protons. Observation of this atom would provide an important new avenue for the study of quantum electrodynamics and quantum chromodynamics.

8

Applications of Atomic, Molecular, and Optical Physics

Atomic, molecular, and optical (AMO) physics lies at a confluence of basic science, applied science, and technology. The applications of AMO physics to the needs of society and to our nation's goals are extensive. They play a conspicuous role in the field, contributing to its vitality and its diversity; they represent a visible return to society for its support of basic science.

From a large list of these applications we have chosen to describe precision measurements, fusion, national security, fiber-optics communication, materials processing, manufacturing with lasers, data-base services, and medicine. This list, however, is far from comprehensive. For lack of space, or because the information may not be publicly available, we have omitted a number of other major activities. Subjects that are entirely omitted, or are only discussed in passing, include environmental monitoring, optical data processing and optical computing, laser isotope separation, inertial confinement, photochemical processing, and laser weapons systems.

PRECISION MEASUREMENT TECHNIQUES

The art of high-precision measurement and the application of precision measurement techniques to basic science and to technology is strongly entrenched in AMO physics. The tradition can be traced to Michelson, who invented the Michelson interferometer to search for the ether drift and realized that he could use it to measure machinists' gauge blocks and length standards to unprecedented precision. The tradition is very much alive today. Within the past decade there have been major advances in precision measurements, the most dramatic of which has culminated in the redefinition of one of the four independent basic units—mass, time, electric current, and length—which are generally considered to make up the cornerstone of physical measurement. Laser metrology became so precise that the accuracy with which the speed of light could be measured was limited by the primary standard of length. As a result, the speed of light was recently fixed by convention, removing the need for an independently defined standard of length. Formerly, the meter was defined as a certain multiple of the wavelength of the red spectral line of krypton; now it is defined as the distance that light travels in a certain fraction of the second. (See Figure 8.1.)

Of all the quantities in physics, time is by far the most accurately measured. The primary time standard in the United States is an atomic clock—basically an atomic-beam magnetic resonance apparatus—located at the National Bureau of Standards in Boulder, Colorado. It has an accuracy of 1 part in 10^{13}, approximately 3 seconds in one million years.

Atomic clocks play an essential role in very-long-baseline interferometry in radio astronomy. Antennas at distant positions in the world observe radio waves emitted by distant radio galaxies and quasars. Hydrogen maser atomic clocks at each antenna synchronize recordings of the phase and amplitude of the signals to a fraction of a microsecond. The recordings are then taken to a central location where the interference patterns formed from the signals are studied. The resultant data correspond to an angular resolution of 0.0001 arc second, far beyond the resolution of an optical telescope. (0.0001 arc second is the angle subtended by 20 centimeters at the distance of the moon.) Atomic clocks are routinely used to synchronize radio and TV communication signals, and they are an essential component of a global system of navigational satellites. The gravitational red shift of time has been measured with a rocketborne atomic clock that was stable to parts in 10^{15} over a period of hours, as described in Chapter 4 in the section on Elementary Atomic Physics. (See also Figure 1.1.)

The ultimate precision of a cesium atomic-beam frequency standard is limited by the brief time the cesium atoms spend in the atomic-beam apparatus. Appreciably longer interaction times have been obtained by using electronic traps to store ions, permitting the creation of clocks of much greater precision. A combination of laser and light radio-frequency field is used to observe the hyperfine transition of the stored ion. A technique called laser cooling (described in Chapter 6 in the section on Laser Spectroscopy) has been successfully applied to reduce the temperature of the trapped ion, an important advantage for precision time measurement. Temperatures in the millikelvin range have been achieved, and still lower temperatures appear to be possible. Not only will the spectral lines be narrower, but the detectability of a small number of ions will be enhanced by their localization in micrometer-size regions. These trapped ions can be used to create an optical frequency standard, that is, an atomic clock operating at an optical frequency rather than a microwave frequency. Neutral atoms have recently been slowed with laser light, even brought to rest in free space. It may be possible to trap these atoms and employ them to create yet another new type of optical atomic clock.

Stable laser sources based on atomic and molecular transitions opened the way to measuring directly the frequency of light in the visible spectrum. (The frequency of light is about 10^6 times higher than the frequency of microwaves.) This feat was performed with nonlinear optical devices used to multiply and compare the signals of a series of lasers operating at successively higher frequencies. The first laser in the chain was related directly to the cesium frequency standard. The measurement provided an absolute reference standard for wavelengths in the visible part of the spectrum; it is this advance that led to the redefinition of the meter described above.

The ability to measure optical frequencies accurately represents an important advance in the transfer of electronic techniques from the radio-wave and microwave regions to optical regions. Advances in optical communications and optical data processing are already emerging as part of this revolution.

Another advance in metrology has been made possible by combining laser interferometry with x-ray diffraction. The technique has been used to measure directly the ratios of wavelengths of selected x-ray transitions to precisely known optical wavelengths. (See Figure 8.2.) These measurements provide calibration lines in the x-ray spectrum and make it possible to use x-ray measurements on heavy muonic atoms to make precision tests of quantum electrodynamics (QED). Furthermore, they have eliminated the need for the X unit formerly used to link x-ray wavelengths to standards units of length.

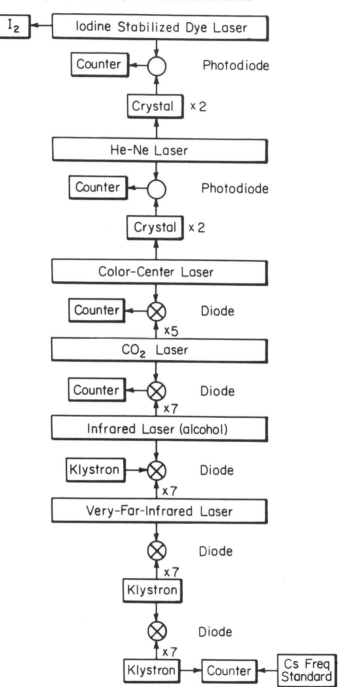

FUSION

The states of reactant material in controlled thermonuclear fusion devices are varied and extreme. Inertially contained plasmas are thousands of times more dense than normal solids; magnetically confined plasmas are a million times less dense than air. Plasma temperatures can reach 10^8 degrees. Atomic and molecular physics provide vital expertise and essential data needed for designing thermonuclear fusion reactors. Fusion research also contributes to atomic physics, for the processes that occur under fusion conditions provide opportunities to observe at close range atomic phenomena that are rarely found on Earth, though they may be common elsewhere in the universe.

Controlled thermonuclear devices burn isotopes of hydrogen to form an ash of helium and neutrons, with a prodigious energy release. (The currently envisioned reaction uses tritium and deuterium as fuel.) In the interior of stars the containment of such a ball of fire is performed effortlessly by gravity, but containment on Earth presents a major challenge. Today's efforts focus on two approaches: magnetic confinement, which uses magnetic fields to curb the escape of charged particles from the plasma, and inertial confinement, in which burning occurs too quickly for the material to escape. Atomic problems are involved in each approach.

FIGURE 8.1 The Frequency of Light and the New Meter. The frequencies of radiowave or microwave signals can be measured quickly and accurately using high-speed electronic methods to count the number of oscillations in a fixed time interval, for instance, 1 second. In the optical regime, however, one normally measures wavelength, using a spectrometer, not frequency. This is because in the past there was no way to count the high-frequency oscillation of light, typically one million times faster than microwave signals. Now this has changed. The drawing shows how the frequency of light has been measured, specifically the frequency of one of the absorption lines in molecular iodine. A series of highly stable lasers of increasing frequency was compared, using nonlinear elements to generate harmonics (exact multiples of the frequency) to "jump" from one frequency to the next. Frequency counters were used to measure the small differences between the multiples of one laser frequency and the frequency of the next. The final laser was locked to a narrow absorption line in molecular iodine. Measuring the frequency of light represents an important advance in metrology, for the frequency of light can be measured much more accurately than its wavelength. One result of this measurement is that the meter has been redefined: the meter is no longer defined as a certain number of wavelengths of a spectral line in krypton; the meter is now defined as the distance that light travels in exactly $1/299,792,458$ of a second. The experiment marks an important step toward transferring electronic techniques to optical frequencies. (Courtesy National Bureau of Standards.)

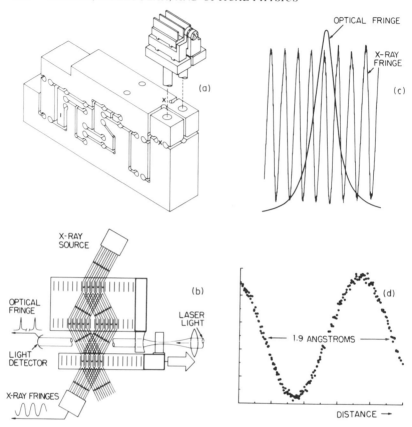

FIGURE 8.2   Measuring the Distance Between Atoms. It is relatively straightforward to measure distance on the scale of 1 meter to 1 part in 10^6 or better. The wavelength of light is less than a millionth of a meter and with modern laser interferometry the number of wavelengths between two surfaces can be counted directly. Accurately measuring distance shorter than the wavelength of light, however, is a much more challenging problem. The distance between atoms in a crystal is less than a thousandth of the wavelength of light, but this distance can be measured to high accuracy using an apparatus that simultaneously combines x-ray and optical interferometry shown in (a). Once calibrated, the crystal can then be used to measure the wavelength of x rays to about the same accuracy.

The heart of the device is an x-ray interferometer made of a single crystal of silicon. (The creation of such crystals is one of the important scientific advances made possible by the semiconductor device industry.) The interferometer, which is composed of three slabs, can be seen at the top of (a). One of the slabs slides laterally, translated by a mechanical stage composed of a series of levers and hinges formed by the holes and slots in the block. Every time it moves by a distance equal to the spacing between atoms in the crystal, the x-ray interference pattern creates a new fringe. The movement is governed by a feedback signal from a laser-driven optical interferometer. The arrangement of the two interferometers is shown in (b). Data showing a single spike from the optical interferometer superimposed on the rapidly varying x-ray fringe are shown in (c). The

Magnetic Confinement

The major problem in magnetic confinement is to insulate the plasma from the vacuum vessel. Bremsstrahlung and line radiation from the capture of electrons into excited states by medium- and high-Z impurity ions, such as iron and tungsten, are important cooling processes, for the plasma is so tenuous that the radiation flows unhindered to the walls. Electron recombination processes play a crucial role in governing this energy loss, for they determine the charge states of the ions. The dominant process is resonant or dielectronic recombination. For tungsten under typical tokamak conditions, for example, at a temperature of 1 keV, if dielectronic recombination is neglected the average charge state is $+50$; when it is included, the charge state is $+28$ and the radiation loss is drastically decreased. Until recently, dielectronic recombination had never been observed directly, and plasma modeling had to rely solely on theoretical rates. Experiments are now possible, and four independent measurements of this process have been made, as described in Chapter 4 in the section on Atomic Dynamics. The results are startling: discrepancies between the experimental values and the theoretical predictions have been as large as a factor of 5. Understanding this process quantitatively is a vital problem for the fusion plasma modeling effort.

Line radiation from impurities in the plasma is a valuable diagnostic probe of the reactant material. Forbidden transitions to the ground states of elements whose ionization energies are near the electron temperature are particularly useful. These lines reveal the ion temperatures by their Doppler profiles and provide ion transport information by their relative line intensities. Unfortunately, the energy levels for highly stripped species are not commonly known: much of the necessary spectroscopic work has had to be carried out on the fusion plasma itself. Such studies are expensive, and they cannot provide the data needed for the next generation of fusion devices. This information can, however, be provided from sources such as laser-produced laboratory plasmas, fast-ion beams, and ion-beam-pumped gases. The experimen-

---

measurement depends on being able to move the interferometer with a resolution much finer than the size of an atom. The enormous resolution of the interferometer is shown by the blowup of one of the x-ray fringes (d). By comparing the x-ray and optical fringe patterns, the distance between the atoms in the silicon crystal can be found directly in terms of the wavelength of light to an accuracy of about 1 part in 10^7. Crystals calibrated by this method can be used to measure x-ray and gamma-ray wavelengths to about the same accuracy. (Courtesy of the National Bureau of Standards.)

tal studies need to be accompanied by theoretical research, particularly on many-electron systems in which relativistic and QED effects are large.

The most successful heating method for several fusion devices is to inject fast neutral atoms of hydrogen, which are ionized and trapped in the reactor. Premature ionization of the neutral beam by collisions with highly stripped impurities on the plasma sheath is a potential problem. New data on electron loss from molecular hydrogen in collisions with highly charged impurity species suggest that early estimates were overly pessimistic and that adequate penetration into the core of today's devices is possible. Little direct data for atomic hydrogen is yet available, however. For the very large machines planned more energetic neutral beams are needed. These beams currently are created by accelerating singly charged ions, H^+ or D^+, to the desired energy and then neutralizing them by charge exchange in a gas such as H_2 or D_2. Unfortunately, at high energy the electron capture cross section is too small for the method to be practical. This problem could be avoided by using high currents of negative ions of light elements. Such ions can easily be stripped of their outer electrons, even at high energies. The physics of single-electron detachment methods for millielectron-volt negative ions remains to be explored.

Hydrogen and light impurities in the center of the plasma are generally fully ionized. However, a neutral hydrogen concentration as low as 1 part in 10^5 can drastically change the ionization state distribution in the plasma and the energy loss due to radiation from impurities. The radiation loss is governed by charge exchange into excited states of the highly stripped impurities. Experimental total cross sections for these processes have become available, but cross sections into particular final states need to be known, and these are so far available only theoretically.

Inertial Confinement

Many of the collision and spectroscopic parameters of concern to magnetic-containment fusion are also important in inertially contained plasmas. One problem that is specific to inertial confinement is development of an efficient driver for compressing the reactant material to ignition conditions. For example, development of a high-powered short-wavelength laser with a well-chosen pulse shape could provide a more efficient driver than existing CO_2 or Nd:glass lasers: the KrF excimer laser is one candidate. An ion beam is a possible alternative to laser light for compressing the reactants. A central question in ion-beam compression is how fast ions stop in material whose density and temperature are high. A vast number of ion-electron and ion-ion collision processes must be understood.

The spectroscopic needs of inertial confinement present an additional complication: the plasma is so dense that it can cause significant changes in the energy-level structure of the ions. For example, shifts of tens of electron volts are expected in the Lyman-alpha radiation from argon. X radiation from the plasma core is the major diagnostic signal that can penetrate the compressed material, but to interpret the signal it is essential to understand the structure of highly relativistic systems, including energy levels, transition rates and fluorescence yields, and satellite structures. Extremely high Z ions occur in ablative pusher targets, which use thin shells of, for example, gold ($Z = 79$), for which QED shifts to the energy-level structure are important. In addition to this spectroscopic information, diagnostic interpretation of the spectra require that the inner-shell ion-ion and ion-electron collision processes, which are responsible for production of the needed inner-shell vacancies, be well understood. The very fast decay of inner-shell vacancies through radiationless transitions provides a femtosecond or even attosecond time scale that makes it possible in principle to utilize these Auger processes to probe the dynamics of the target compression. A thorough theoretical understanding of the de-excitation of deep vacancies in multiply ionized atoms is required for this purpose. Although much information on ion-atom collisions has become available over the past decade, experimental work on inner-shell vacancy production in ion-ion collisions is rare.

Considering all these problems, it is evident that the fusion energy program will continue to demand intensive efforts from atomic physics.

NATIONAL SECURITY

AMO physics contributes to the national security by providing basic information and devices that are vital to our defense systems. The challenge of accurate navigation offers one example. Atomic clocks and frequency standards are at the heart of our modern navigational systems. The accuracy of these devices is prodigious—clocks with an accuracy approaching parts in 10^{15} have been developed, and even higher accuracy appears to be possible. Atomic clocks are used in a precise positioning system that makes it possible to determine where one is anywhere on Earth to within 10 meters. The system can be used for civil as well as military navigation. The creation of the atomic clocks and frequency standards that make this system possible dramatizes the interplay between basic and applied science in atomic physics. These devices grew out of basic research in spectroscopy and the structure of matter. In addition to their role in navigation, they are crucial to the operation of secure communications systems, and they have applications in areas of science such as very-long-baseline interferometry, determination of the fundamental constants, and tests of relativity.

Atomic clocks in use include the hydrogen maser, the cesium atomic beam clock, and the optically pumped rubidium cell. The precision of these devices is ultimately limited by the second-order Doppler effect, the time dilation due to the motion of the atoms. To overcome this barrier, the motion of the atoms must be eliminated. Stored ion spectroscopic techniques provide one solution. During the last 5 years enormous strides have been made in designing ion traps and in cooling trapped ions to a small fraction of a kelvin. Recently a beryllium-ion frequency standard has been operated within a stability that is within a factor of 3 of high-performance cesium-beam devices. Future work using mercury ions is expected to improve this by several factors of 10.

Another technique that can reduce the second-order Doppler effect is the cooling and trapping of neutral atoms. An atomic beam of sodium has been cooled to 4 percent of its original velocity. This advance has catalyzed interest in atom traps, for if the atoms are sufficiently slow they can be trapped and stored in a neutral-particle trap.

The problem of communicating with submarines provides a second example of how AMO research can play an important role in the national defense. Our nuclear fleet is the least vulnerable of all of our defense systems. Communicating with the submarines, however, is difficult because radio waves are strongly absorbed by seawater. Laser communication is a promising technique. Ocean water has a spectral window in the blue-green region. A number of laser systems are being studied for generating blue-green light for this application and also for underwater surveillance, for illumination, and for bathymetry. The excimer lasers, which are described in Chapter 6 in the section on Lasers—The Revolution Continues are particularly promising. Because these lasers operate in the violet and ultraviolet, their light must be downshifted to the blue-green. Several atomic gases are promising candidates, including barium, bismuth, and lead. These vapors, however, require high temperatures, and they are corrosive. Raman-shifting techniques using the molecules hydrogen and deuterium provide a second approach for shifting the light from the ultraviolet to the blue-green. Work continues toward developing a system that satisfies all the requirements of wavelength, efficiency, power, and lifetime.

A highly sensitive frequency-selective detector is needed for most of the applications of the blue-green laser. One proposed device is based on atomic absorption and fluorescence. A narrow-band, wide-field-of-view detector can be created by enclosing an atomic gas between two filters. The first filter transmits the to-be-detected radiation, while the second filter blocks it. The light excites a resonance transition in the gas. The fluorescence occurs at a different wavelength, which is transmitted by the second filter. Very high quantum efficiency has recently been achieved with this scheme.

Highly sensitive frequency-selective detectors have military applications in the optical, infrared, and millimeter-wave regions. One scheme for a millimeter/submillimeter detector exploits the high absorption cross sections of highly excited Rydberg atoms for very-long-wavelength radiation. A pair of Rydberg states is chosen to be resonant with the incident radiation. By putting the system in an electric field so that the higher Rydberg state is field ionized while the lower is not, one produces, on absorption of a photon, a free electron, which is readily detected. The system behaves like a phototube for millimeter waves and microwaves, with the additional advantage of being highly frequency selective.

The high power and high efficiency of carbon dioxide lasers developed over the past decade suggest the potential for application for high-resolution radar. Unfortunately, their radiation at a wavelength of 10 micrometers can be attenuated under certain atmospheric conditions. Many atmospheric windows exist in the millimeter and submillimeter regions, however, and although efficient lasers are not available at these frequencies, the radiation can be generated by pumping various organic molecules with a CO_2 laser.

The free-electron laser has been demonstrated in the past decade. This device is potentially capable of generating highly efficient, tunable short-wavelength radiation. Radiation at 1.5 micrometers has been generated with almost 10 watts of power, and the device has been operated in the visible. Chemical lasers are another class of high-power lasers of molecular interest. These, too, are quite efficient, though restricted with respect to wavelength.

High-power laser, microwave, and particle-beam devices require high-power switches that can operate at rates greater than 1 kilohertz, pass more than 10 kiloamps, and hold off more than 50 kilovolts. Spark-gap and diffuse discharge switches are two possible gaseous devices. There is no difficulty in closing such switches rapidly, but opening the switch rapidly, which requires removing more than 10^{15} electrons/cm^3, is extremely difficult. One possible technique employs laser-driven chemical reactions. The idea is to populate an excited state of a species that has a high electron attachment or recombination cross section. Current proposals along this line require very high power in order to open the switch, but if a process could be found in which the photons act as a ''catalyst,'' this could be a valuable approach.

These are but a few examples of the contributions of AMO physics to military technology. Numerous other devices could be cited, including laser surveillance systems, fiber-optics communications, and optical-processing technology. In addition, the results of basic AMO research are often essential for understanding military scenarios, for instance cloud formation, nucleation phenomena, ionospheric disturbances, the evolution of atmospheric species, and aircraft signatures.

AMO physics contributes to our national security system in one further respect: it trains doctoral-level physicists who are capable of carrying forward the research and development programs in national and industrial laboratories. The skills that AMO physicists acquire in atomic, molecular, and electronic processes; in optics and lasers; and in advanced experimental and theoretical techniques are vital to these programs.

FIBER-OPTICS COMMUNICATIONS

In fiber-optics communication light pulses representing digital information are launched from an electrically driven light source into a specially prepared glass fiber and are detected at the distant end and reconverted into electrical signals. High-capacity information can often be provided more economically by fiber optics than by radio, coaxial cable, or satellites. The advance is profoundly changing communications. For example, hitherto long-distance digital voice transmission has been hampered by the great bandwidth required. This problem is disappearing as fiber optics, with their vast bandwidth and with its ready adaption to digital signals, becomes pervasive throughout the world.

The creation of the laser first sparked serious interest in communicating large amounts of information by means of light beams. Glass fibers appeared attractive for the transmission or medium, but only through arduous research in the physics of light in fibers, and the chemistry of fiber composition and fabrication, could the medium become practical. The first problem was the loss due to absorption of near-infrared radiation by the glass. The loss in conventional glass is prohibitive, it is measured in thousands of decibels per kilometer; as a result of chemical purification, the loss in glass has been reduced to 0.1 to 1 dB/km. As a result, the separation of repeater stations in a transmission system can be up to 100-200 km. (A high-capacity coaxial system requires repeaters every 1 to 2 km.) Other improvements in the glass were required to limit the dispersion in the fiber (the degree to which a sharp pulse of light entering a fiber becomes spread out in time as it travels along the fiber) and to assure that the light does not scatter from the surface. As a result of close interactions between physicists, chemists, and engineers, industrial techniques have been devised to produce fibers with properties that a few years ago would have seemed almost beyond belief. A new industry has been created.

Fiber-optics communications requires light sources that are on a physical scale with the hairlike fibers. The requirements are demanding; they must provide high intensity, be reliable, work at the required wavelengths, be reasonably efficient in power conversion, and be relatively inexpensive. Semiconductor diode lasers meet these requirements. In order that the generated light be confined in an optical cavity, elaborate structures are built up involving semiconductor regions of varying band gaps and with varying dopants. To achieve efficient radiative recombination of the holes and electrons, and to assure that the devices are reliable, techniques needed to be developed to achieve high crystallographic quality even though the devices include regions of widely different chemical composition. A great deal of development work went into perfecting new crystal growth processes, leading to the creation of an important new branch of the semiconductor industry.

Research in fiber optics holds the promise of providing a whole new field of integrated-optics devices. It may be possible to do much of the signal processing that is now done electrically by processing the optical signals themselves rather than first converting them into electrical signals. This day may be rather far off, but there is no doubt that optical elements such as switches, polarizers, isolators, or wavelength division multiplex devices will be important parts of fiber-optic communication systems in the near future.

The fiber-optic communication industry is large and growing rapidly. Worldwide sales in 1984 may approach $1 billion, and in subsequent years greater volumes are expected. The industry is highly competitive on an international as well as on a national scale. The industry is completely dependent on fundamental research in physics and chemistry: a healthy climate for fundamental research is essential for any nation that hopes to compete in this increasingly important industry.

MANUFACTURING WITH LASERS

Industrial applications for lasers were evident when the first ruby laser punched a hole through metal. The earliest applications were all highly specialized, but lasers are now also employed for numerous routine manufacturing tasks. (See Figure 2.2.) Their use is spreading rapidly, and lasers can be expected to play major roles in wide areas of the U.S. industrial enterprise in years to come.

Foremost among the advantages of lasers for materials processing and manufacturing are these:

—Lasers can deliver energy at far greater density than is possible by any other technology. The ability to achieve high temperature in short times permits laser processing of almost any material and opens the way to new machining, welding, annealing, heat treating, and chemical procedures.

—Laser light is effectively inertialess and it can be focused with optical precision. As a result, lasers are ideally adapted to automatic control techniques, to robotics, and to high-speed processing of intricate shapes. Thus, lasers are expected to play an increasingly important role in the development of new flexible manufacturing plants.

Laser Drilling: Lasers are used to drill both exotic and ordinary material. Tasks include highly specialized applications such as drilling diamond-wire pulling dies and creating multitudes of air-cooling holes in jet engines and routine jobs such as for drilling holes in baby-bottle nipples. Laser drilling is practical for all the metals, plus ceramics, gemstones, semiconductors, plastics, wood, and rubber.

Laser Cutting: Laser cutting employs the light to heat the workpiece to its melting temperature and a jet of gas to remove the vaporized or molten material. The focused laser beam acts as an effective point tool, a tool that never makes physical contact with the material. Applications range from cutting out suit patterns to cutting sheet-metal parts. In addition to textiles and metals, the list of materials that can be cut includes ceramics, quartz, glass, composites and exotic aerospace materials, plastics and fiber-reinforced plastics, wood, leather, and fiberboard.

Laser Welding: Because laser light can be accurately controlled and directed, laser welding is a direct competitor to electron-beam welding. However, laser welding has an important advantage. It does not require the workpiece to be under vacuum; room conditions are sufficient. Most applications include metals and metal alloys, though some nonmetals can be welded. In some cases, it is practical to weld dissimilar material that could not otherwise be joined. One-half-inch steel plate is routinely welded; even thicker materials can sometimes be handled.

Other Applications: Lasers are used in the semiconductor industry for annealing, resistor trimming, and silicon scrubbing. Lasers are used for marking, soldering, and surface treatments including hardening, cladding, and alloying. For example, one factory that manufactures power-steering units currently uses 20 laser systems to heat treat the guiding surfaces.

The Future: The factory of the future is expected to be highly automated. The laser, which lends itself naturally to operation under control systems and to automation, provides an ideal technology. The Japanese government has recently initiated a $60 million, 5-year joint university/industry program to study flexible manufacturing systems with lasers, including the integration of lasers with control systems. Approximately 6000 lasers are currently used in factories around the world. Altogether more than 1000 lasers were sold for material processing in 1982, with prices ranging from $5000 to $500,000. The market for laser material-processing equipment has been growing at 20 percent per year over the past several years, and this rate is expected to be maintained over the next decade. In 1982, the sales were $110 million; in 1992 they are expected to be $700 million. At the end of the decade lasers will be commonplace in manufacturing plants ranging from small shops to heavy industrial plants and automated factories. The economic benefit to the United States from the use of lasers in manufacturing and materials processing is expected to be enormous: one survey estimates that lasers will lead to the creation of 600,000 new jobs in these areas (*Newsweek*, Vol. 108, p. 78, October 18, 1982).

The question of manpower is a potentially serious problem for the orderly development of laser-based manufacturing in the United States. There is already a shortage of qualified personnel in the optics and electro-optics industries. As discussed in Chapter 2, the total production in the United States of physicists with Ph.D. degrees in optics is less than 50 a year; the production at the bachelor's level is also small. Unless this problem is successfully addressed, our manufacturing capability may be limited by a personnel shortage.

MATERIALS PROCESSING

Laser-Induced Surface Chemistry

Laser-induced surface chemistry provides new processing techniques with applications in semiconductor electronics and electro-optics. Submicrometer features have been produced in a single processing step, and novel microstructures have been made.

In one process, a focused laser beam drives a small-scale chemical reaction at the surface, either in the gas or liquid phase. Submicrometer patterns can be produced by doping, etching, and deposition—all based on laser-controlled chemistry. The deposition rate can be much larger than possible by any other means. In a second process, radiation from a high-power laser creates radicals in gas-phase parent molecules that then react at the surface. The exact reaction can be specified by selecting the proper laser wavelength. Since the dissociation occurs in the gas phase, substrate heating is negligible.

Lasers can be used to dope surfaces and to etch them. Typically, a pulsed excimer laser is used to photodissociate a gas-phase compound and simultaneously to melt a nearby substrate. A small amount of doping can be incorporated with each pulse. Both Si and GaAs have been doped using this technique.

Dry etching of dielectrics, semiconductors, and polymers can be achieved by photodissociating methyl-containing compounds. By illuminating solid polymers with high-intensity, pulsed, UV light, photochemical decomposition of the polymer cross-linking can cause well-defined surface etching.

Ion Implantation

The bulk properties of solids can be altered or completely transformed by the introduction of small amounts of impurity atoms. Accelerator-based atomic physics has created techniques for introducing these impurities in highly novel fashions. Using a small accelerator, ion-implantation methods allow impurities to be added free from the usual material constraints of diffusion and solid solubility. Ion bombardment can create electrically insulating layers or new surface alloys; ion-induced defects can induce material diffusion or ion-induced segregation. Near-surface modification by ion bombardment has produced amorphous metals and led to the creation of surfaces that are tough and relatively free from corrosion and oxidation. Ion-implantation methods have been applied to the epitaxial growth and doping of diamonds and to the creation of ion-implanted solar cells.

Ion- and molecular-beam techniques are now used for the production of integrated circuits. Ion beams can be controlled so precisely that it is possible to construct integrated circuits directly on a silicon wafer. Using molecular-beam epitaxy, solids can be produced in layers of precisely controlled thickness. The layers can be only a few atoms thick.

High-power lasers provide a unique tool for annealing the surface of solids and providing diffusion in a limited region. A very short pulse-probe laser can be used to monitor the state of the surface and study the dynamics of the melt.

DATA-BASE SERVICES

AMO physics provides atomic and molecular data that are essential to wide areas of science and technology and to our nation's energy, military, and environmental programs. An enormous growth of raw data has been made possible by advances in laser spectroscopy, in neutral- and charged-particle scattering, and in theoretical and computational methods. To be useful, it is essential that the data be accurately evaluated, systematically compiled, and efficiently disseminated. Here is a summary of some of the applications.

Laser Physics and Development: Electron and photon collision cross sections and potential energy curves are essential for modeling laser plasmas and designing high-power lasers. Atomic-energy levels, wavelengths, transition probabilities, and atomic lifetimes are needed to model lasing action and to assess probable population inversions. Recombination cross-section data are required for the design of gas-discharge lasers.

Nuclear Fusion Energy: Requirements include collision rates and cross sections for electron impact excitation, ionization, and dielectronic recombination; atomic-energy levels and wavelengths for identification of impurity elements in fusion plasmas, for plasma diagnostics and modeling, and for calculating plasma-cooling effects due to radiation from impurity atoms; photon mass attenuation coefficients and electron stopping powers are needed to design reactor blankets.

Isotope Separation: Atomic-energy levels, transition probabilities, and lifetimes are all needed in order to design a laser isotope-separation process.

Atmospheric Monitoring: Photon and electron interaction data are essential for understanding such problems as radio-wave propagation in the ionosphere, the theory of the aurora, cosmic-ray-induced cascade showers and carbon-14 production in the atmosphere. Molecular spectral data are needed for remote sensing of the complex chemistry of the polluted atmosphere.

Medical Physics: Photon attenuation data are required for cancer therapy using accelerator and isotopic sources of radiation and for the imaging studies and dosimetry of internal nuclear medicine.

Industrial Applications: The needs are broad. To cite two examples: electron and photon attenuation and cross-section data are required to determine the effectiveness of ionizing radiation for food processing, the sterilization of medical supplies, and chemical processing such as the polymerization of plastics; atomic-energy levels, lifetimes, and transition probabilities are needed for laser development, surface-property studies, and the design of lamps for street lighting and other applications.

National Security: Atomic transition probability data are essential for modeling nuclear explosions in the atmosphere; photon attenuation data are needed to design shielding against ionizing radiation from nuclear weapons. Atomic-energy level, lifetime, and transition probability data are all needed for x-ray laser development.

Astrophysics: The needs include extensive electron excitation, ionization, and recombination data on astrophysical ions. Wavelengths, atomic-energy levels, and transition probabilities are needed for identifying spectral lines and determining the elemental composition of astrophysical sources.

Most atomic and molecular data centers in the United States are associated with the National Standard Reference Data System, managed by the Office of Standard Reference Data at the National Bureau of Standards. This program develops important data bases of physical and chemical properties needed by industry, academia, and government. The program coordinates the activities of 23 continuing Data Centers and 31 shorter-term Projects.

Each major Data Center monitors an important disciplinary area and develops and maintains a major data base that is subsequently made available in published, computer-readable, and on-line formats. The Projects answer the need for specialized data bases in particularly important areas. The scientists associated with the Data Centers and

Project monitor the literature in their fields of expertise, compile and evaluate data, mathematically predict data in difficult to measure regions, and prepare major compilations and data bases for the U.S. technical community.

Computer-based data networks can now be created. Such networks are essential to meet the nation's growing needs for AMO data and data from other disciplines.

MEDICAL PHYSICS

AMO physics contributes to the basic science of medicine and to the creation of new techniques for medical practice. We describe here two activities: laser surgery and NMR body imaging.

Laser Surgery

Laser surgery is becoming a standard practice as lasers replace scalpels in more and more surgical procedures. Fields include general and cardiovascular surgery, urology, dermatology, dentistry, plastic surgery, and oral surgery. A recent conference on medical and surgical application of lasers attracted 1300 physicians and surgeons from 31 countries.

One type of cervical cancer that formerly required a hysterectomy can now be treated in a doctor's office or on an outpatient basis using CO_2 laser light. A simple, painless procedure provides a high cure rate for a condition that otherwise demands a major operation. The magic of the laser light is in its being able to remove the malignant cells without disturbing the underlying tissue. Lasers are used by eye surgeons for procedures that include repair of a detached retina, treatment of diabetic retinopathy, glaucoma, and iridectomy. All of these procedures capitalize on the ability to control and deliver the laser's energy with extremely high precision. Treatment of diabetic retinopathy, for example, involves destroying cells on the outer layer of the retina without damaging the underlying layer. Iridectomy requires making a tiny hole in the iris, which the laser can do without damaging any of the nearby tissue.

Laser surgery is intrinsically sterile. The radiation tends to cauterize a wound, reducing the bleeding and scarring. Optical fibers can deliver the light precisely where it is needed. For example, fatty deposits that clog coronary arteries can be vaporized by laser light delivered through a fiber-optic endoscope. The method holds the promise of revolutionizing coronary arterial surgery.

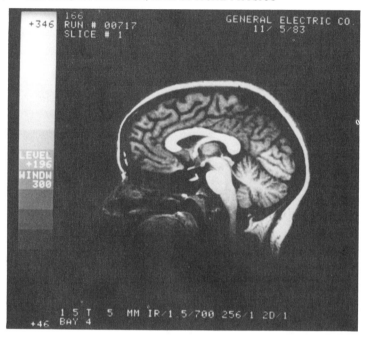

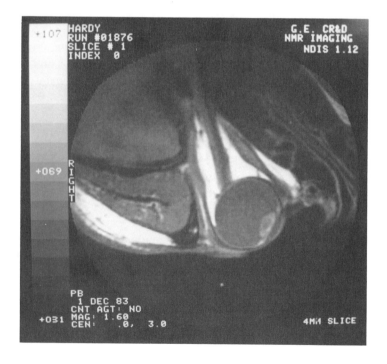

The wide range of wavelengths available from different lasers is an important asset to laser surgery. Infrared radiation from CO_2 lasers is most commonly used to remove tissue. The radiation is absorbed rapidly by water and is effective in coagulating blood vessels. Eye surgery generally employs green light from argon-ion lasers, since this is transmitted freely by the lens and the aqueous humor. Argon-ion laser light is also used to treat crimson birthmarks: the light passes through the skin and is absorbed by the underlying blood capillaries that are responsible for the unsightly stain and congeals them.

In certain applications, laser light can be used to induce therapeutic photochemical reactions. For instance, tumors can be treated by injecting the patients with the chemical HPD, which is selectively retained by malignant cells. Red light from a tunable dye laser excites the HPD, which transfers the excitation to the cell, killing it. The technique provides a new field of therapy for cancer.

Magnetic-Resonance Whole-Body Imaging

Using nuclear magnetic resonance (NMR) it is now possible to peer into the human body as gently and clearly as we view its surface. (See Figure 8.3.) Magnetic-resonance imaging (MRI) is nonperturbing, noninvasive, and free of ionizing radiation. It can achieve spatial resolution of 1 mm and can often clearly distinguish different types of tissue. As far as is known, the method is without biological hazard. Recent trials using first-generation equipment have demonstrated MRI to be useful for detecting diseases in the brain, spinal column, heart, thorax, abdomen, liver, and kidneys. MRI has aroused the intense

---

FIGURE 8.3 Magnetic Resonance Imaging. Nuclear magnetic resonance provides a new way to form images of the body's interior using a noninvasive procedure that is believed to be without hazard. In contrast to x rays, magnetic resonance imaging can display the differences between soft tissues. Tumors or lesions and differences between diseased and healthy tissue can often be identified precisely and rapidly. The upper picture shows a cross-sectional image through the midplane of the head. The spinal nerve is clearly visible. The lower image (facing right) is through a parallel plane passing through the eye. The orbit lens and optic nerve are easily discerned. The data collection time for each of these images was slightly over 3 minutes. Magnetic resonance imaging is expected to have a major impact on medical diagnosis. It is made possible by contributions from many fields: magnetic resonance from AMO physics, tomographic techniques developed by mathematicians and medical physicists, and superconducting magnets from low-temperature research. Minicomputers and modern data-processing techniques also play a vital role. (Courtesy of General Electric Research and Development, Schenectady, New York.)

interest of both the medical community and the commercial sector: some practitioners believe that it provides the biggest single advance in medical diagnosis since the discovery of x rays.

Nuclear magnetic resonance spectroscopy has been used in chemistry and biochemistry laboratories for over 30 years as an analytical tool to study the conformation and dynamics of biological molecules. Within the last 5 years two new variants of the MRI technique have emerged: proton imaging, which makes it possible to produce cross-sectional pictures of the human body, and phosphorous imaging, which makes it possible to study physiological function in vivo over volumes of 50 cm^3. For example, phosphorous imaging can distinguish between healthy and unhealthy tissue, providing information that can be crucial to cardiologists and peripheral vascular physicians. Applications for NMR body imaging include the following.

Thorax: Blood produces a strong NMR signal, which reveals the major vessels and the cardiac chambers. NMR can be used to assess tissue perfusion, to measure cardiac function, and to examine the myocardium (i.e., the middle and the thickest layer of the heart wall composed of muscle). The absence of NMR signals from air-filled lungs makes it easy to observe a hemorrhagic lung, which might be caused by pulmonary embolus, pleural effusion, or tumor.

Abdomen: NMR is superior to all the previous methods for diagnosing lesions of the liver and for detecting many liver diseases. Cystic lesions in the urinary tract are easily differentiated from solid lesions. Aneurisms—blood-filled sacs formed by the dilation of the walls of the abdominal aorta—are clearly demonstrated. The presence of arteriosclerosis with degenerative changes in the wall of the descending aorta can be observed.

Pelvis: NMR scanning of female patients shows the extent of malignant new growth that infiltrates the ovary, uterus, and cervix. In males, it can provide a definitive diagnosis of malignant growth before it invades through the prostate capsule.

Musculo-Skeletal: Observations in patients with rheumatoid arthritis have shown that the presence of inflammation can be imaged, making it possible to assess the response of the inflammation to antiinflammatory drug therapy. Imaging can also reveal how a tumor shrinks as a result of chemotherapy.

Companies in the medical equipment business have been quick to grasp the commercial potential of MRI technology. At least 15 manufacturers are said to have imaging systems available for controlled clinical research or under development. As of 1983, it was estimated that investigational MRI systems were in use in 35 U.S. sites and 12 sites abroad. The average cost of this equipment is about $1.3 million per unit. A total of about $800 million of internal funds has been spent on MRI systems development by 13 member companies of the National Electrical Manufacturers Association. The market for MRI systems by the year 1990 is assessed at about $2 billion to $3 billion.

The development of MRI whole-body imaging required advances from many areas of science. Understanding of the magnetic resonance properties of nuclei in various media comes from studies in physical chemistry and biology; the superconducting magnets, which are essential parts of the imaging apparatus, are a product of low-temperature materials research. Minicomputers and modern data-processing techniques also play an essential role. The basic idea, however, magnetic resonance, is due to Felix Bloch at Stanford University and Edward Purcell at Harvard University, who developed NMR in 1946 in order to understand the motions of nuclei in matter. It is difficult to think of an argument more eloquent than this for the benefit to mankind of basic research in science.

Further Reading

Proceedings of the International Conference on Atomic Physics
ATOMIC PHYSICS 9, Seattle, 1984; R. S. VanDyck, Jr., and E. N. Fortson, eds. World Scientific Publishing Co., 1985.
ATOMIC PHYSICS 8, Goteburg, 1982; I. Lindgren, A. Rosen, and S. Svanberg, eds. Plenum Press, New York, 1983.
ATOMIC PHYSICS 7, Cambridge, Massachusetts, 1980; D. Kleppner and F. M. Pipkin, eds. Plenum Press, New York, 1981.

Proceedings of the International Conference on the Physics of Electronic and Atomic Collisions
ELECTRONIC AND ATOMIC COLLISIONS, invited papers at the 13th International Conference, Berlin, 1983; J. Eichler, I. V. Hertel, and N. Stolterfoht, eds. North-Holland, Amsterdam, 1984.
PHYSICS OF ELECTRONIC AND ATOMIC COLLISIONS, invited papers at the 12th International Conference, Gatlinberg, Tennessee, 1981; S. Datz, ed. North-Holland, Amsterdam, 1982.

Proceedings of the International Conference on Laser Spectroscopy
LASER SPECTROSCOPY VI, Interlochen, Switzerland, 1983, H. P. Weber and W. Luthy, eds. Springer-Verlag, Berlin, 1984.
LASER SPECTROSCOPY V, Alberta, Canada, 1981; A. R. W. McKellar, T. Oka, and B. P. Stoicheff, eds. Springer-Verlag, Berlin, 1981.

Proceedings of the Second International Conference on Precision Measurement and Fundamental Constants, Gaithersburg, Maryland, 1981
PRECISION MEASUREMENT AND FUNDAMENTAL CONSTANTS, II, B. N. Taylor and W. D. Phillips, eds. Natl. Bur. Stand. (U.S.) Spec. Publ. 617 (1984).

ADVANCES IN ATOMIC AND MOLECULAR PHYSICS
D. Bates and B. Bederson, eds. Academic Press, New York. Volume 16 (1980)-Volume 20 (1984).

PROGRESS IN ATOMIC SPECTROSCOPY
Parts C and D, H. J. Beyer and H. Kleinpoppen, eds. Plenum Press, New York, 1983.
Parts A and B, W. Hanle and H. Kleinpoppen, eds. Plenum Press, New York, 1979.

175

Index